Reactivity and Mechanism in Organic Chemistry

2nd Edition

Reactivity and Mechanism in Organic Chemistry

2nd Edition

By

Hendrik Zipse
Ludwig-Maximilians-Universität München, Germany
Email: zipse@cup.uni-muenchen.de

ROYAL SOCIETY
OF **CHEMISTRY**

Print ISBN: 978-1-83916-743-0
EPUB ISBN: 978-1-83916-754-6

A catalogue record for this book is available from the British Library

The Royal Society of Chemistry is a charity, registered in England and Wales, Number 207890, and a company incorporated in England by Royal Charter (Registered No. RC000524), registered office: Burlington House, Piccadilly, London W1J 0BA, UK, Telephone: +44 (0) 20 7437 8656.

Visit our website at www.rsc.org/books

Printed in the United Kingdom by CPI Group (UK) Ltd, Croydon, CR0 4YY, UK

Preface

The current text is intended to provide an introduction to the quantitative description of organic reactivity and is aimed at the student enrolled in a chemistry bachelors or masters program. In the first chapter we develop qualitative molecular orbital theory as an important tool for the description of bonding phenomena. The combination of molecular orbital theory and thermochemical data is then employed to rationalize concepts such as hyperconjugation, the anomeric effect, the *gauche* effect, molecular strain and aromaticity. A second chapter introduces transition state theory as a general framework for the discussion of organic reactivity phenomena, and also illustrates its relationship to potential energy surfaces and simple rate equations. On this basis we then develop more specific reactivity concepts commonly used in organic chemistry textbooks, such as the Bell–Evans–Polanyi principle, Marcus theory, the HSAB principle, Hammett correlations, the Mayr–Patz equation and frontier molecular orbital theory. A short description of the inner workings of solvent effects is also included here. How these reactivity models are applied is demonstrated for pericyclic reactions and selected rearrangement reactions, for reactions involving transient intermediates, such as radicals, diradicals or carbocations, and for reactions involving classical electrophile/nucleophile combinations. I would like to thank many colleagues worldwide for their valuable comments on the first edition of this text.

Hendrik Zipse

Reactivity and Mechanism in Organic Chemistry, 2nd Edition
By Hendrik Zipse
© Hendrik Zipse 2023
Published by the Royal Society of Chemistry, www.rsc.org

Contents

Reactivity and Mechanism in Organic Chemistry, 2nd Edition
By Hendrik Zipse
© Hendrik Zipse 2023
Published by the Royal Society of Chemistry, www.rsc.org

2 Reactivity Models in Organic Chemistry 54

3 Pericyclic Reactions 101

Subject Index 190

1 Structure and Bonding in Organic Molecules

1.1 Atomic Orbitals as Building Blocks

Bonding phenomena in organic molecules are often described in terms of **valence bond (VB)** or **molecular orbital (MO)** theory. Valence bond theory is based on localized electronic configurations (VB configurations), in which valence electrons occupy defined positions in the overall system. Combination of several of these configurations yields valence bond states, which define the ground and excited state properties of the system. The qualitative description of VB configurations through Lewis structures is the primary tool used in organic chemistry for the description of organic molecules. One of the weak areas of VB theory concerns molecules with extended π-systems, and this is also where MO theory has particular strengths. The basic tenet of MO theory is that the overall wavefunction of the system can be described by a product of one-electron functions called orbitals. While quantitative (or *ab initio*) versions of MO theory are highly successful in describing (and thus predicting) the properties of organic molecules, the focus here is on the qualitative application of MO theory by using **atomic orbitals** (**AOs**) as building blocks for the construction of MOs. This approach is often termed **linear combination of atomic orbitals (LCAO)** and is easily understood with respect to the atomic orbitals of the carbon atom in its electronic ground state (Figure 1.1).

Reactivity and Mechanism in Organic Chemistry, 2nd Edition
By Hendrik Zipse
© Hendrik Zipse 2023
Published by the Royal Society of Chemistry, www.rsc.org

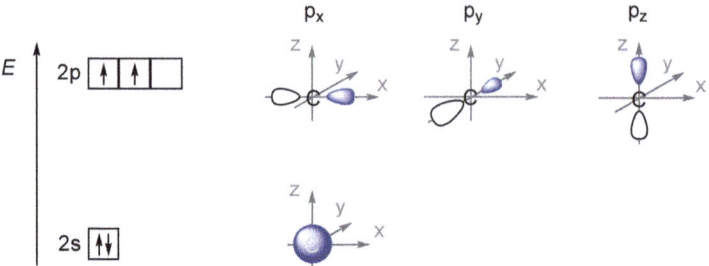

Figure 1.1 Atomic orbitals of the carbon atom in its electronic ground state.

We concentrate on the valence space orbitals here and thus neglect the (energetically low lying) 1s atomic orbital holding another two electrons. The schematic description in Figure 1.1 already implies four basic rules of quantum chemistry that also apply to molecular orbitals.

1. Electrons located in a given orbital are characterized by a defined energy and a defined probability density. The latter defines the shape of the orbital and is often visualized pictorially by selecting a particular cut-off value for the density.
2. The Pauli exclusion principle (or **Pauli principle**) states that orbitals can be occupied by only one electron. A second electron is permitted under the condition of having opposite spin to the first one.
3. The **Aufbau principle** states that electrons are added to the atomic and molecular orbitals bottom-up: that is, filling the lowest energy orbital first and then adding electrons to the next higher orbitals until all electrons have been placed.
4. **Hund's rule**, although valid explicitly only for atoms, also applies to most molecular systems and states that filling up orbitals of equal energy (that is, degenerate orbitals) deposits one electron in each orbital first, and then starts adding a second electron to the singly filled orbitals (until all electrons are placed).

1.2 Construction of Molecular Orbitals

The construction of molecular orbitals from atomic orbitals follows simple rules that can be described as follows.

1. The combination of *n* atomic orbitals generates *n* molecular orbitals. In the most simple case two atomic orbitals interact to generate two molecular orbitals. One of these molecular orbitals

is **bonding** in nature and thus energetically more favorable than the contributing atomic orbitals, the second one is **antibonding** and energetically less favorable. For H_2, as a simple example, the two 1s atomic orbitals holding one electron each combine to yield the energetically lower (bonding) MO 01, in which no change of phase (no nodal plane) occurs along the line connecting the two hydrogen atoms. In the higher lying (antibonding) MO 02 a change of phase occurs along this bond path.

Quantitative evaluation of the bonding situation in H_2 indicates that the energy lowering of the bonding MO relative to the contributing AOs (ΔE_1) is always smaller than the energy increase of the antibonding combination (ΔE_2). In the example of H_2 chosen here, calculations at B3LYP/6-31G(d) level yield orbital energies of -0.434 Hartree for the favorable MO and $+0.100$ Hartree for the unfavorable MO, which have to be compared to -0.316 Hartree for the 1s atomic orbital in the hydrogen atom. The H_2 molecule is bound because two electrons occupy the bonding orbital, while the antibonding MO 02 remains empty. The quantum mechanically calculated structure of this latter orbital is largely similar to that derived by qualitative MO theory. This is not so for the lower lying MO 01, where the contributing atomic orbital building blocks are combined such that a single ellipsoid molecular orbital results (Figure 1.2b).

2. The combination of atomic orbitals into molecular orbitals is only possible under the condition that both AOs share common symmetry properties. Relevant symmetry elements can in this case be the rotational axis running through the two interacting atoms or mirror planes in which the interacting atoms are located. Selecting the rotational axis running through the two interacting atoms as an example, one set of p-type atomic

---

Common energy units for orbital energies and their conversion factors

$$1 \text{ Hartree} = 27.21 \text{ eV} = 627.51 \text{ kcal mol}^{-1} = 2625.5 \text{ kJ mol}^{-1}$$
$$1 \text{ eV} = 23.06 \text{ kcal mol}^{-1} = 96.5 \text{ kJ mol}^{-1}$$
$$1 \text{ kcal mol}^{-1} = 4.184 \text{ kJ mol}^{-1}$$

The Hartree (sometimes also referred to as the atomic unit (a.u.) of energy) is the most common energy unit used in quantum chemical calculations, while the electron volt (eV) is often used to report results from spectroscopic measurements.

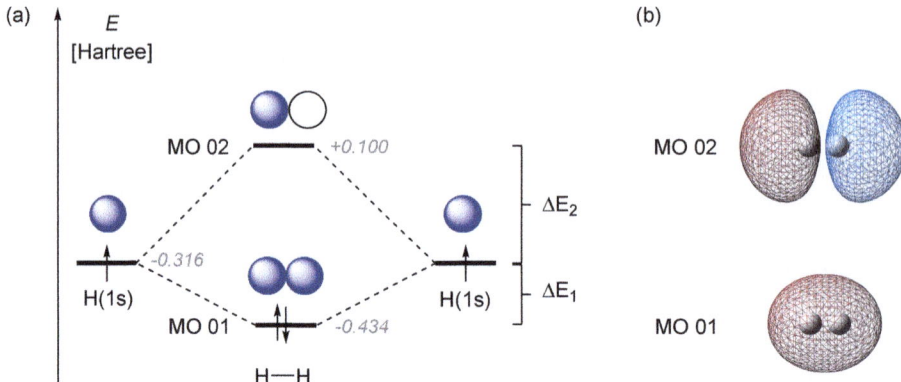

Figure 1.2 (a) Bonding in H_2 as described by qualitative MO theory. (b) MOs for H_2 calculated quantum mechanically (B3LYP/6-31G(d) theory).

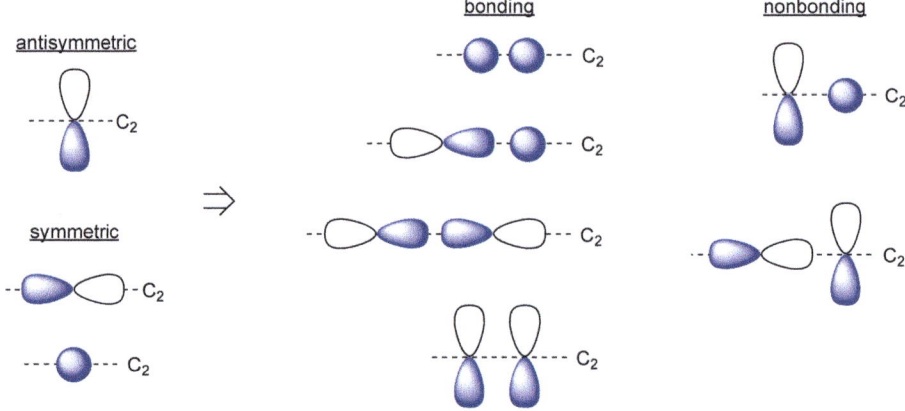

Figure 1.3 Bonding and non-bonding interactions between atomic orbitals.

orbitals are **antisymmetric** with respect to a twofold axis running through the interacting atoms, while other p-type as well as all s-type AOs are **symmetric** with respect to the same symmetry axis (Figure 1.3). Bonding interactions subsequently arise only between AOs sharing the same symmetry properties with respect to the common C_2 axis, while all other interactions are non-bonding.

3. The magnitude of (atomic) orbital interactions depends on their **overlap.** This is quantified by the overlap integral (often denoted as S) and depends on the relative orientation of the AOs as well as their distance. Orbitals that are p-type overlap better when oriented in the longitudinal orientation, as depicted in Figure 1.4a,

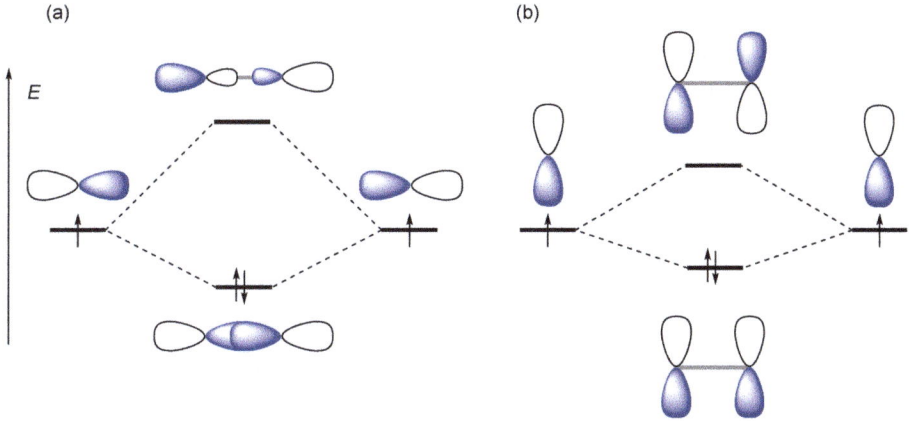

Figure 1.4   Interaction of p-type atomic orbitals arranged in (a) longitudinal and (b) lateral fashion.

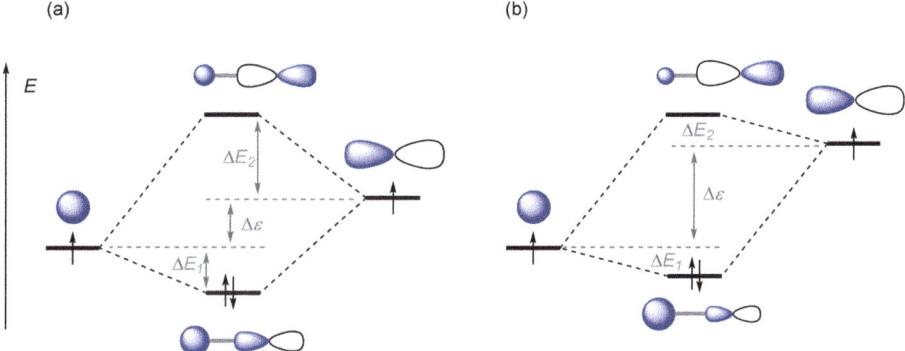

Figure 1.5   Interaction of s- and p-type atomic orbitals with (a) a small and (b) a large initial energy gap $\Delta\varepsilon$.

compared with a situation involving lateral arrangement, as in Figure 1.4b. For first-row elements, such as carbon, this difference is the cause for the higher bond strengths of σ- as opposed to π-bonds.

4. The magnitude of the interaction ΔE between orbitals of different energy is inversely proportional to their energy difference $\Delta\varepsilon$. For the example of the interaction of an s- and a p-type orbital of similar energy (implying a small energy gap, $\Delta\varepsilon$) the situation is shown in Figure 1.5a. The energy changes resulting from orbital interactions ΔE_1 and ΔE_2 are in this case significant, which is in contrast to the situation of a larger energy gap as depicted in Figure 1.5b. Please also observe that the shape of the resulting

molecular orbital is more strongly influenced by the energetically closer lying atomic orbital. In the example chosen here the energetically lower lying molecular orbital thus has higher contributions from the (lower lying) s-type atomic orbital, while the less favorable MO has higher contributions from the (higher lying) p-type AO. These differences in MO structure increase with increasingly larger energy separation of the contributing atomic orbitals. In a semiquantitative way these trends can also be expressed as $\Delta E \sim S^2/\Delta\varepsilon$.

For very large energy separations, as is the case for core and valence space orbitals, this implies vanishingly small interaction energies and thus provides the basis for the conceptual separation of electrons into the core and the valence regions.

All of the above concepts can be seen at work when analyzing the bonding situation in hydrogen fluoride (HF). In this case valence space orbital interactions involve, on the hydrogen side, the 1s atomic orbital located at −13.6 eV. The fluorine valence space orbitals include the three (isoenergetic) p-type AOs at −18.7 eV and the much lower lying 2s AO at −46.4 eV (Figure 1.6). In order to identify possible orbital interactions by symmetry type, it is helpful to orient the H−F molecule in a Cartesian coordinate system such that its molecular axis aligns with one of the principal axes. For this we select here the *x*-axis, which also coincides with the C_2 axis differentiating the symmetry properties of the three p-type AOs of fluorine. Choosing this orientation,

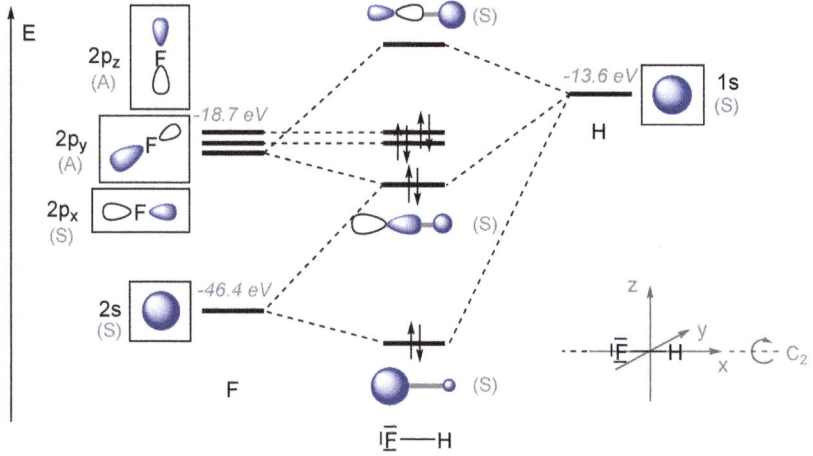

Figure 1.6   Schematic orbital interaction diagram for the valence space of hydrogen fluoride (HF, not drawn to scale).

the fluorine p_y and p_z AOs are antisymmetric with respect to rotation around the C_2-axis, while the p_x and the 2s AOs are symmetric (as is the hydrogen 1s AO). On comparing the symmetry properties of AOs on the hydrogen and fluorine side, it becomes immediately obvious that the antisymmetric fluorine p_y and p_z AOs have no interaction partner on the hydrogen side and thus remain unchanged on formation of the HF molecule. The fluorine 2s and $2p_x$ AOs and the hydrogen 1s AO then interact to generate three new molecular orbitals, which inherit the symmetry properties of the contributing AOs. The structure of the lowest lying of these three MOs is dominated by the energetically lowest lying AO (the fluorine 2s AO). The second highest MO has notable contributions from both the $2p_x$ AOs and the hydrogen 1s AO, as has the highest lying third MO. Filling the four valence space electron pairs into the available MOs bottom-up fills the lowest two of the symmetric MOs as well as the two unchanged fluorine p-type AOs with two electrons each. This implies that the F–H single bond is mainly described by the second of the symmetric MOs and that the three non-bonding electron pairs drawn in the Lewis structure are actually quite different in character: two of these correspond to the unchanged fluorine p-type AOs, while the third is the lowest lying symmetric MO mainly composed of the fluorine 2s AO.

1.3 Orbitals of π-Bonded Organic Molecules

1.3.1 π-Systems of Linear Polyenes

The prototype for an organic molecule with a π-system is ethylene. All atoms of ethylene are located in a plane horizontal to the principal axis running through the center of the C–C double bond (Figure 1.7). Adding two more C_2 axes horizontal to the principal axis and two additional mirror planes implies that ethylene belongs to the D_{2h} point group. For our discussion here we will concentrate on the horizontal mirror plane (σ_h) as a symmetry element for the valence space atomic orbitals. Sorting through the hydrogen and carbon atomic orbitals with an eye to their symmetry properties, it is easily seen that almost all AOs are symmetric with respect to the principal symmetry plane. This is true for the hydrogen 1s and the carbon 2s AOs, and also for the $2p_x$ and $2p_y$ AOs located at both carbon atoms (assuming the orientation of ethylene as depicted in Figure 1.7). The only antisymmetric atomic orbitals in this system are the two carbon $2p_z$ AOs, which combine to form the two molecular orbitals of the π-system. The more numerous symmetric AOs

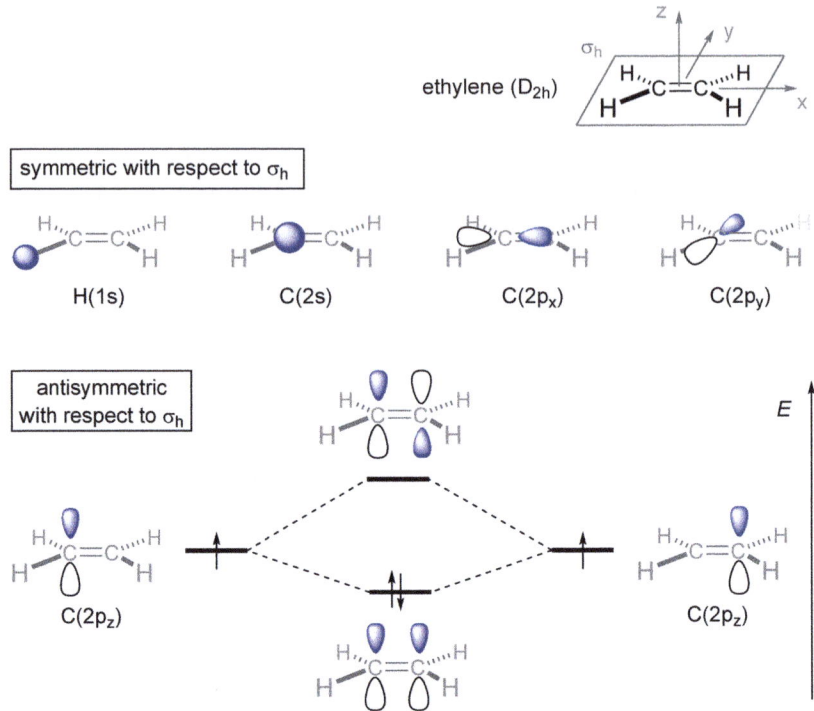

Figure 1.7 Bonding situation in ethylene.

combine to form the (more extensive) molecular orbitals for the σ-bonding system. The two MOs of the π-system differ in that the lower lying MO contains no phase change when moving from the left to the right side of the π-bond, while a phase change does occur in the higher lying MO. The two electrons available for the π-type MOs are filled into the energetically lower lying one, leaving the higher lying π-MO empty.

The allyl cation (Figure 1.8) shares with ethylene the characteristic that all atoms are located in a common mirror plane. This again allows for the separation of valence space atomic orbitals into a symmetric set (building up the σ-bonding system) and an antisymmetric set. This latter group is composed of three carbon p-type AOs combined into three π-type molecular orbitals. In building the three MOs it is practical to use the two ethylene π-MOs as building blocks for the two outer positions of the allyl system. These "stretched ethylene" MOs are quite similar to those of ethylene and differ only in their smaller energy separation due to the larger distance of the contributing two C(2p) AOs. In order to see how these two fragments interact with the

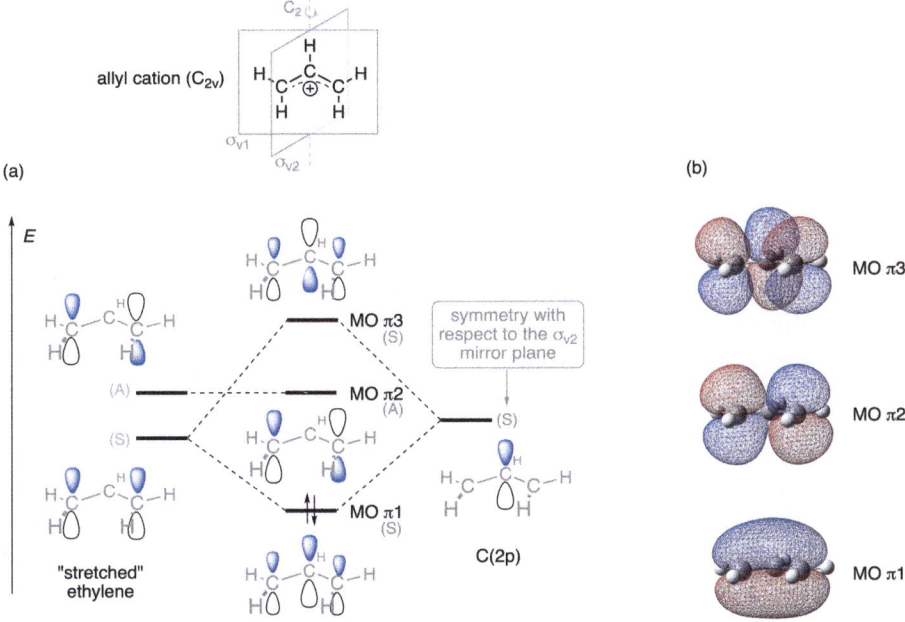

Figure 1.8   Molecular orbitals for the allyl cation π-system when (a) derived from qualitative MO theory or (b) calculated quantum mechanically (B3LYP/6-31G(d) theory).

C(2p) AO located at the central allyl carbon atom we use the vertical σ_{v2} plane dividing the allyl cation in the middle as an additional symmetry element.

The C(2p) AO located at the central allyl carbon and the energetically lower lying of the two "stretched ethylene" fragment MOs are symmetric with respect to this symmetry plane, while the higher lying "stretched ethylene" fragment MO is antisymmetric. Combining the two symmetric building blocks from the left and right side we generate one energetically favorable and one less favorable combination, termed MO π1 and MO π3 (Figure 1.8). In the latter MO the orbital phases continuously change from center to center, while there is no phase change in the former. The antisymmetric "stretched ethylene" fragment MO remains unchanged (in shape or energy) and is thus identical to the third molecular orbital of the π-system, termed MO π2. We complete construction of the allyl cation π-system by filling the two π-electrons into the lowest lying MO π1. How these pen-and-paper derived predictions compare to molecular orbitals calculated quantum mechanically can be seen in Figure 1.8b, where we see the three allyl cation π-MOs calculated with the B3LYP/6-31G(d) method frequently employed for molecular systems. While the two higher lying π-MOs are conceptually quite similar to our qualitative derivations, the lowest lying MO π1 shows only one single orbital phase extending over the three carbon atoms. Similar to what we have seen previously for the hydrogen molecule, the three individual p-type AOs visible in our qualitative picture of MO π1 thus merge into a single phase whenever no phase change occurs. This will also be the case for the larger π-systems discussed below, but we will nevertheless continue to use the qualitative MO presentations with individual p-type AOs for the sake of simplicity. The construction of the π-systems of allyl radical and allyl anion is, in principle, identical to the procedure chosen here for the allyl cation, except for the larger number of π-electrons used in the last step. Quantitative studies show, however, that the energies of the π-orbitals increase with an increasing number of π-electrons, that is, absolute orbital energies are higher (less negative) in the allyl radical than in the allyl cation. Extending the allyl π-system by one more center leads to buta-1,3-diene. In the *anti* conformation shown in Figure 1.9 this C_{2h}-symmetric system again has all atoms located in one mirror plane and displays a C_2 axis running through the central C–C bond. The four C(2p) atomic orbitals antisymmetric to the horizontal mirror plane again combine to yield the four MOs of the π-system. The strategy of using "stretched ethylene" building blocks for the outer two centers (carbons C1 and C4) as well as the inner two centers (C2

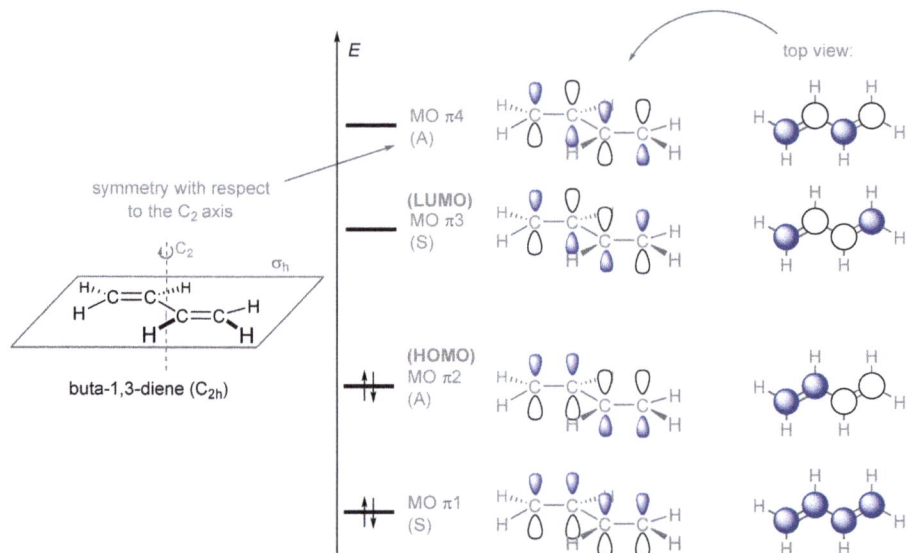

Figure 1.9 Molecular orbitals for the π-system of buta-1,3-diene.

and C3) can also be used here in combination with the C_2 axis as the differentiating symmetry element. This latter symmetry element is also the differentiating property of the four resulting molecular orbitals of the π-system, as shown in Figure 1.9. The energetically lowest of the π-MOs shows no phase change along the π-system and is symmetric with respect to rotation around the C_2 axis. Moving upwards in energy we can observe that the symmetry of the π-MOs changes regularly between being antisymmetric (A) and symmetric (S). We also note that the number of phase changes (sometimes also referred to as *nodal planes*) increases by one from MO to MO. The second lowest MO π2 must therefore display one phase change and be antisymmetric with respect to the C_2 axis. These two properties can only be satisfied with the AO combination shown for MO π2 in Figure 1.9. We complete construction of the buta-1,3-diene π-system by adding four π-electrons to the lowest two π-MOs. This makes MO π2 the **highest occupied molecular orbital** (**HOMO**) and MO π3 the **lowest unoccupied molecular orbital** (**LUMO**). These two orbitals are often considered to be the most relevant MOs for determining the reactivity of π-systems (see Chapter 2, Section 2.7 on frontier molecular orbital theory).

The π-MOs of conjugated linear π-systems can actually be constructed with respect to the three observations we just made for buta-1,3-diene.

1. The lowest π-MO shows no phase change along the perimeter of the π-system.
2. The number of phase changes (nodal planes) along the perimeter of the π-system increases by one on moving from one π-MO to the next higher one.
3. The symmetry of the π-MOs changes regularly from one π-MO to the next higher one.

The application of these three rules can be demonstrated for the penta-2,4-dien-1-yl radical shown in Figure 1.10. In its all-*trans* conformation all atoms of this system are again located in one mirror plane and, after additional consideration of a second mirror plane and one C_2 axis, we can recognize this molecule as C_{2v} symmetric. This is not necessarily evident from the single Lewis structure shown here, but combination with the other two resonance structures obtained by moving around π-electrons and the unpaired spin generates the properly symmetric description. The π-system extends over five carbon atoms in this case and we will thus combine five C(2p) atomic orbitals into five π-MOs. For the sake of simplicity we will draw the

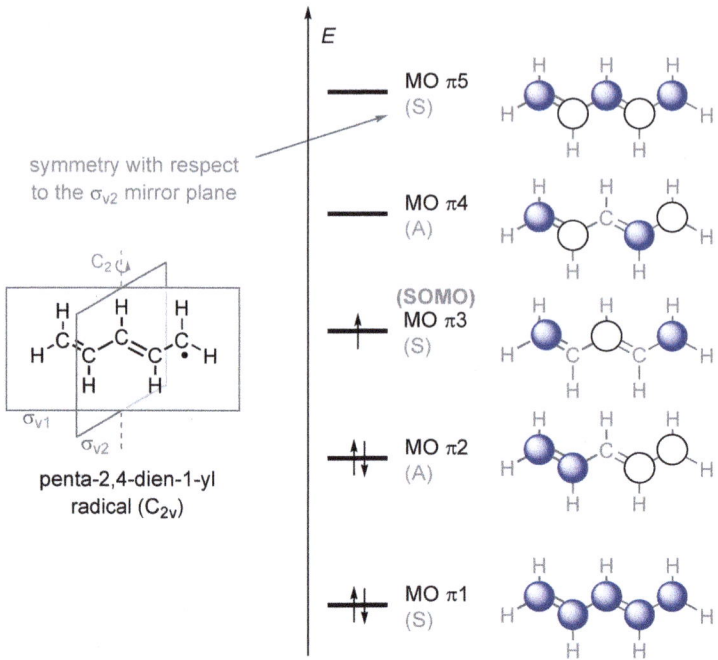

Figure 1.10   Molecular orbitals for the π-system of the penta-2,4-dien-1-yl radical.

system as seen from the top and the contributing C(2p) atomic orbitals will thus be seen as filled or empty circles. The lowest lying MO π1 again has no change in phase along the perimeter of the π-system and is overall symmetric with respect to reflection through the σ_{v2} plane bisecting the molecule. The next MO π2 must therefore be antisymmetric and display one phase change. These conditions are most easily met by the MO structure shown in Figure 1.10, where the central carbon atom C3 has no contribution to MO π2. This property is quite typical for MOs of π-systems with an uneven number of centers and has already been seen for MO π2 of the allyl cation in Figure 1.8. Continuing upwards, MO π3 must be symmetric with respect to the σ_{v2} plane and possess two phase changes. These conditions are met by the MO structure shown in Figure 1.10, where every second carbon atom makes no contribution to the MO, while the phase at the other carbons changes sign regularly. The remaining two orbitals can be constructed following the same logic. In the highest lying MO π5 the orbital phases change regularly from one center to the next one, a property seen in the highest lying MOs of all linear π-systems. Adding five electrons to these five MOs bottom-up we see that the unpaired electron ends up in MO $\pi 3$, which is therefore termed the **singly occupied molecular orbital (SOMO)** of the system.

The MOs of larger linear π-systems can be derived in very much the same way as discussed above for the smaller system. Comparison with quantum mechanically calculated MOs indicates no difference in relative MO energies as a function of nodal planes, but shows that MO coefficients vary more strongly than we have anticipated up to now. This becomes increasingly more important with increasing system size. The qualitative MOs in Figure 1.11 have been drawn such as to reflect these differences.

The bonding situation in molecules containing C–C triple bonds is conceptually very similar to those containing C–C double bonds in that two p-type atomic orbitals located on the participating carbon atoms combine into a set of bonding and antibonding π-orbitals. In contrast to ethylene, this type of π-orbital formation occurs twice in acetylene using p-type atomic orbitals of different symmetry. Using the orientation of the molecule depicted in Figure 1.12 with the principal axis of the molecule oriented in the x direction of the Cartesian coordinate system, a first set of π-orbitals is constructed using the

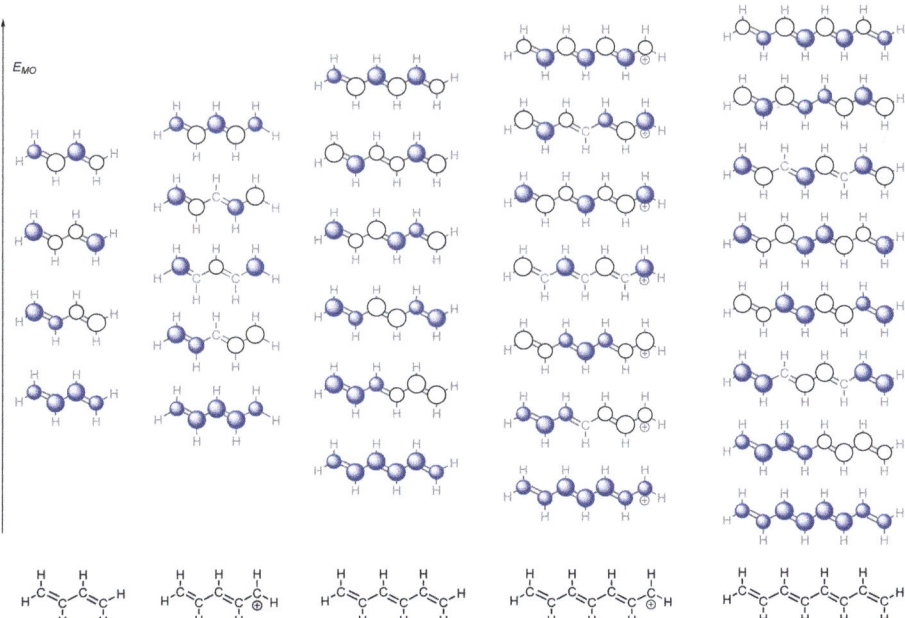

Figure 1.11 Molecular orbitals for the π-systems (left to right) of buta-1,3-diene, penta-2,4-dienyl cation, hexa-1,3,5-triene, hepta-2,4,6-trienyl cation and octa-1,3,5,7-tetraene.

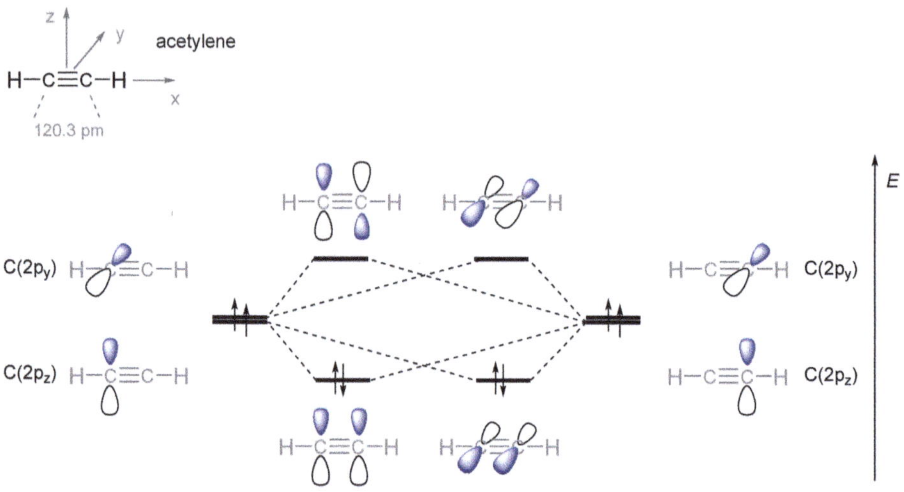

Figure 1.12 π-Molecular orbitals for acetylene (C–C distance from ref. 1).

C(2p$_z$) AOs and a second set using the C(2p$_y$) AOs. This gives rise to two doubly occupied π-orbitals with identical orbital energies but different spatial orientation. Moreover, a quantitative comparison to the π-system in ethylene shows that the orbital energies of the occupied π-orbitals in acetylene are lower compared with those in ethylene. This is due to the shorter C–C bond distance and thus slightly larger lateral overlap in acetylene (120.3 pm) compared with ethylene (132.9 pm).

1.3.2 π-Systems of Cyclic Molecules

The construction of the molecular orbitals of cyclic π-systems is most readily achieved by starting from predefined building blocks instead of individual C(2p) atomic orbitals. This is easily demonstrated when deriving the π-MOs of quadratic cyclobutadiene (CB, Figure 1.13a). This hypothetical molecule has attracted considerable attention as the prototype of an anti-aromatic molecule. Assuming the highest possible (D_{4h}) symmetry, the π-system of CB can be constructed using the π-MOs of ethylene as building blocks, once for the upper and once for the lower half of CB, as shown in Figure 1.13. Using the vertical mirror plane designated as "σ_v", the two ethylene π-MOs can be characterized as symmetric (S) or antisymmetric (A) with respect to this mirror plane. The two symmetric ethylene MOs can be combined into two new π-MOs, the lower of which has no phase change along the perimeter of the π-system. The energetically less favorable combination (termed MO $\pi2$) has two phase changes and is isoenergetic

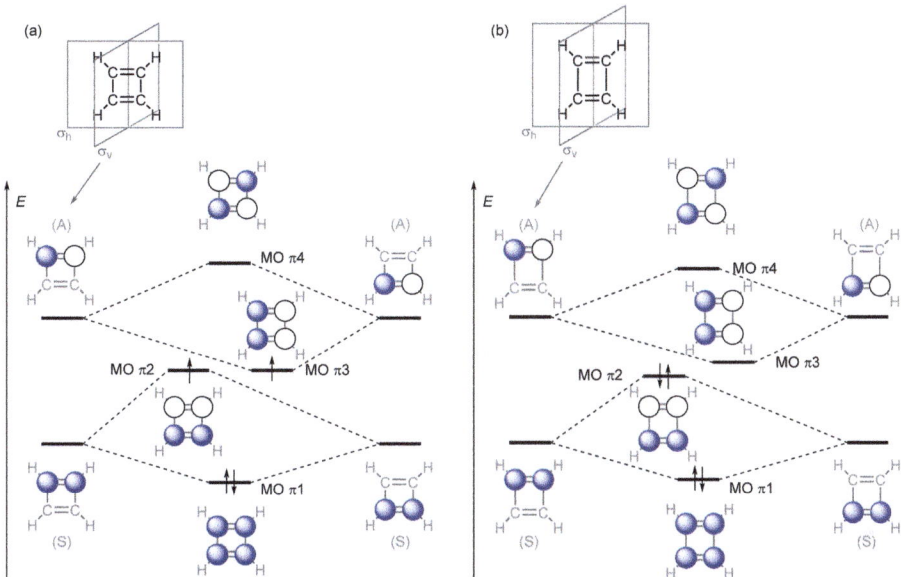

Figure 1.13 Molecular orbitals for the π-system of (a) quadratically planar and (b) rectangular cyclobutadiene (CB).

with MO π3 obtained through combination of two antisymmetric ethylene MOs. The highest lying MO π4 also derives from these latter two building blocks, but is characterized by four phase changes along the perimeter of the π-system. The molecular orbitals π2 and π3 of quadratically planar CB are isoenergetic as is readily seen in Figure 1.13, where a simple 90° rotation of π2 generates π3. When filling the four π-MOs of quadratically planar CB with four π-electrons, we realize that the two isoenergetic (or degenerate) MOs π2 and π3 are filled with only one spin-up electron each, at least when attempting to satisfy Hund's rule. This leads to an unfavorable triplet ground state of quadratically planar CB and provides a driving force for the distortion to rectangular (but still planar) cyclobutadiene. The consequences of this distortion are seen in the orbital interaction diagram in Figure 1.13b, which still involves the same ethylene building blocks as before. However, the overlap between these building blocks is smaller in rectangular CB, which leads to a smaller energy splitting between the newly formed CB π-MOs (*e.g.* MO(π2)−MO(π1) in quadratically planar or in rectangular cyclobutadiene). As a consequence of this deformation, the π-MOs π2 and π3 are not isoenergetic anymore and the four π-electrons can be placed spin-paired into the lower two molecular orbitals. Due to the still rather small energy separation of π2 and π3, rectangular

CB has a small singlet/triplet energy gap and shows exceedingly high reactivity in a variety of bimolecular reactions.

The π-type molecular orbitals of benzene as the prototype of aromatic molecules can also be constructed most easily using fragment MOs from other π-systems. As shown in Figure 1.14 the most straightforward strategy involves a combination of "stretched" cyclobutadiene MOs with "stretched" ethylene MOs. How these fragments combine into the six π-MOs of benzene is guided by the fragment symmetry properties with respect to the two vertical mirror planes labeled as "σ_{v1}" and "σ_{v2}". Comparison of the symmetry labels for the CB fragment orbitals on the left side of the MO interaction diagram with the ethylene fragment MOs on the right side quickly shows that the two MOs with (AS) or (AA) symmetry on the left have no interaction partner of like symmetry on the right. These two fragment MOs will thus remain unchanged from their original structure when building the benzene π-MOs. Fragment MOs exist on both sides of the diagram for the (SS) and the (SA) symmetry types, and these can therefore be combined to construct new benzene π-MOs. The lowest lying MO π1 is again fully bonding and symmetric with respect to the two symmetry planes. The next two MOs, π2 and π3, show two phase changes and are isoenergetic, as are MOs π4 and π5 with four phase changes. The least

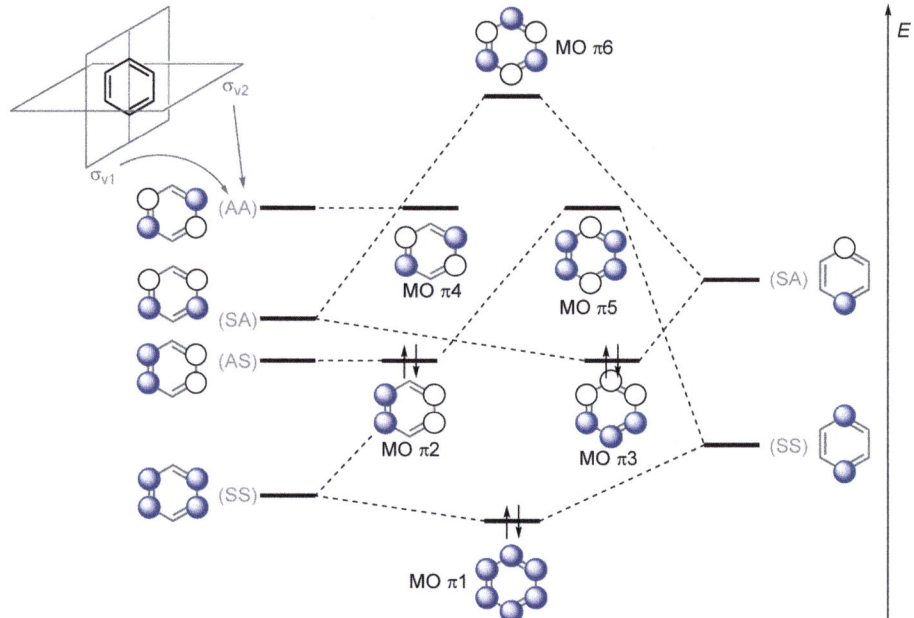

Figure 1.14 Molecular orbitals for the π-system of benzene in its hexagonal structure.

favorable MO π6 is again characterized by the largest number of nodal planes, that is, the orbital phase changes from one carbon to the next. Filling the benzene π-MOs with six electrons, we now see that we can fill the isoenergetic MOs π2 and π3 with four electrons, thus avoiding the unfavorable high-spin situation encountered in quadratically planar cyclobutadiene.

A comparison of these two π-systems also points to the origin of **Hückel's rule**, according to which π-systems with $(4n + 2)$ π-electrons display **aromatic** character. This rule, with $n = 0, 1 ...$, simply defines the number of electrons needed to fully occupy isoenergetic orbitals of cyclic π-systems, with benzene being the best known case for $n = 1$. π-Systems with $(4n)$ π-electrons are often termed **anti-aromatic**, and cyclobutadiene is a typical case for $n = 1$. The π-MOs of cyclobutadiene and benzene shown in Figures 1.13 and 1.14 also illustrate that the lowest π-MO of cyclic systems is always fully binding (shows no phase change), and that, on moving up in energy, the number of nodal planes increments by two. The isoenergetic MOs π2 and π3 thus both have two phase changes on moving along the perimeter of the π-system, while this number increases to four for MOs π4 and π5.

The orbital energies of cyclic π-systems can also be derived using the **Frost–Musulin diagram**. As shown in Figure 1.15 for some typical

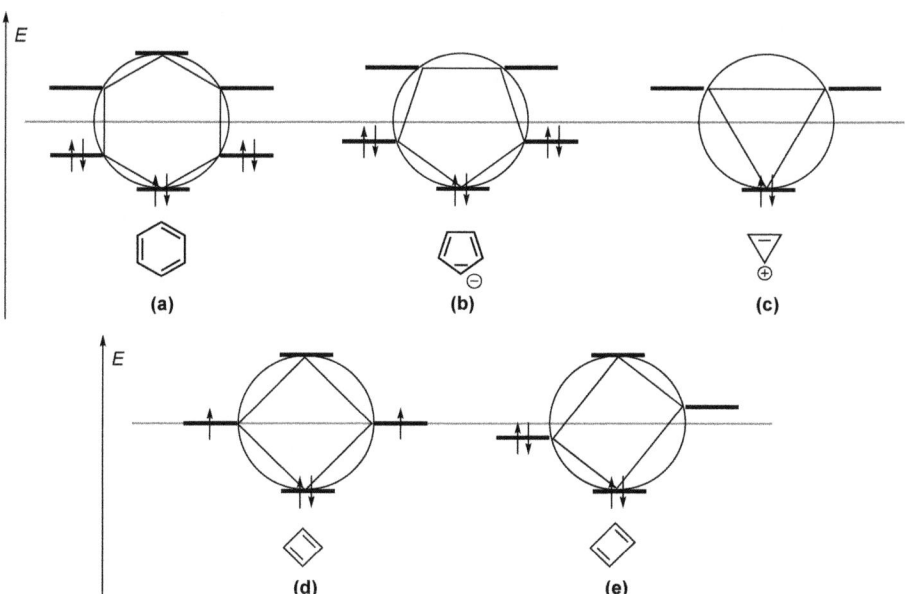

Figure 1.15 Frost–Musulin diagrams for (a) benzene, (b) cyclopentadienyl anion, (c) cyclopropyl cation, (d) quadratically planar cyclobutadiene and (e) rectangular cyclobutadiene.

cases, this diagram is constructed by inscribing the polygon defining the outer perimeter of the π-system into a circle such that the polygon's vertices are all positioned on the circle and one of the vertices coincides with the lowest point of the circle. The horizontal (grey) line running through the center of the circle defines the energy of non-interacting C(2p) atomic orbitals and all vertices of the polygon define the orbital energies of the π-system relative to this reference.

This quickly developed diagram again illustrates the Hückel ($4n + 2$) rule, in a pictorial manner, and predicts "aromatic" character for benzene, cyclopentadienyl anion and cyclopropyl cation in that isoenergetic MOs are either fully filled or empty. For quadratically planar cyclobutadiene the MOs π2 and π3 are again predicted to be isoenergetic and of equal energy to the non-interacting C(2p) AOs used as building blocks. In combination with the four π-electrons this leads to a triplet ground state for this system. The consequences of distorting cyclobutadiene to a rectangular shape can also be followed in the Frost–Musulin diagram as long as we keep one vertex pointing down and one pointing up. Under these conditions the remaining two vertices intersect with the circle such that two energetically different π-MOs are predicted, the lower of which takes up two of the π-electrons.

1.3.3   π-Systems Containing Electronegative Heteroatoms

Adding electronegative elements such as oxygen or nitrogen to one of the π-systems discussed above changes the shape and energies of the respective π-molecular orbitals in a characteristic way. This can readily be demonstrated for formaldehyde, whose π-system is closely related to that of ethylene. We recall for the latter that the two π-MOs are constructed from two C(2p) atomic orbitals in a bonding and an antibonding fashion, thus generating the π-MOs π1 and π2, as shown in Figure 1.16.

In formaldehyde one of the C(2p) atomic orbital building blocks is replaced by an O(2p) atomic orbital, whose orbital energy is much lower at −15.9 eV compared to the carbon analog (−10.7 eV). This difference in orbital energies leads to a larger contribution of the O(2p) AO to the lower lying MO π1 and a smaller contribution to the higher lying π2. Also, the energies of the formaldehyde MOs π1 and π2 are lower than those in ethylene, the effect being larger for π1 than for π2. Similar observations can be made for many other heteroatom-containing π-systems such that: (1) introduction of electronegative elements lowers the energy of the respective molecular orbitals; and (2) the structure of π-MOs is more strongly influenced by energetically close-lying AOs compared to energetically more remote AOs. These principles are also

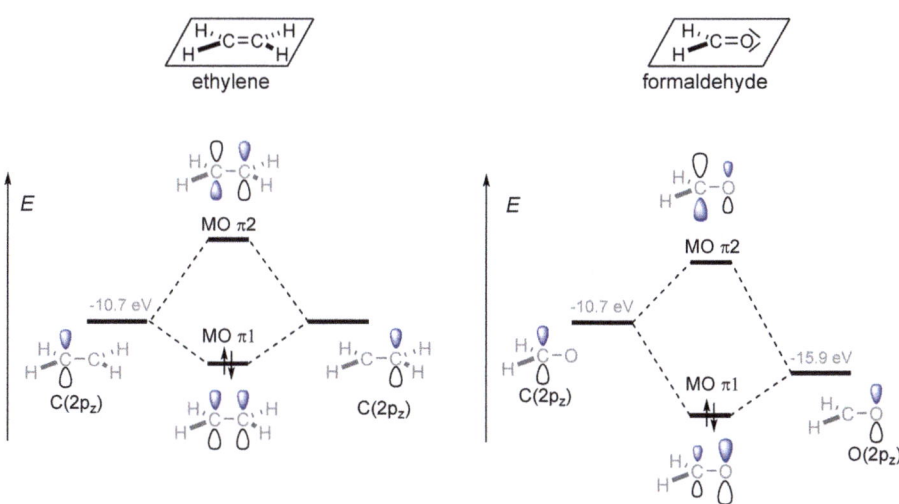

Figure 1.16 Molecular orbitals for the π-systems in ethylene (left) and form-aldehyde (right).

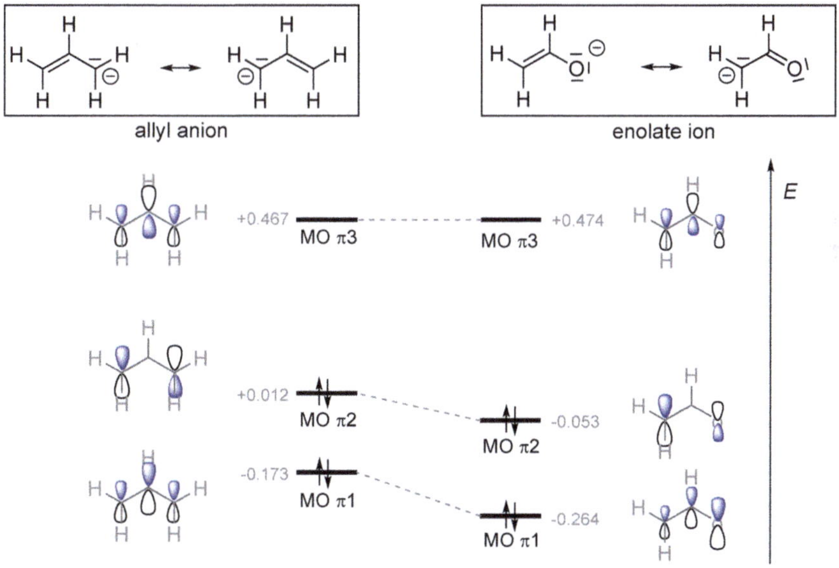

Figure 1.17 Molecular orbitals and orbital energies (in Hartree) for the π-systems in allyl anion (left) and enolate ion (right).

at work in the π-system of the enolate ion, whose π-orbitals are occupied by four π-electrons and closely related to those of the allyl anion (Figure 1.17). In the allyl system the two outside carbon atoms have identical contributions in all MOs due to the symmetry of the system. Formal replacement of one of these carbons by oxygen leads to the enolate ion, where the lowest of the three π-MOs is strongly polarized towards the oxygen atom and lowered by 0.091 Hartree relative to the allyl anion π1 orbital (using orbital energies calculated at RHF/6-31G(d) level). The orbital lowering is somewhat less significant for the π2 orbital, with −0.065 Hartree, where the carbon atom now has a larger contribution. The highest lying π3 orbital shows practically no change in orbital energy and is dominated by the two carbon atoms.

The π-systems in buta-1,3-diene and acrolein (Figure 1.18, energies drawn to scale) also differ in that formal replacement of a terminal methylene group in buta-1,3-diene by oxygen leads to acrolein. The consequences of this formal replacement can again be seen in the orbital energies, as well as the structures of the four π-orbitals. The lowest of these (MO π1) is strongly polarized towards the oxygen atom and significantly lower in energy compared with the MO π1 in butadiene. This can, to a lesser extent, also be stated for the π2 molecular orbital. Only a small energy lowering can be noted for π3, which carries its largest coefficient at the terminal carbon atom and

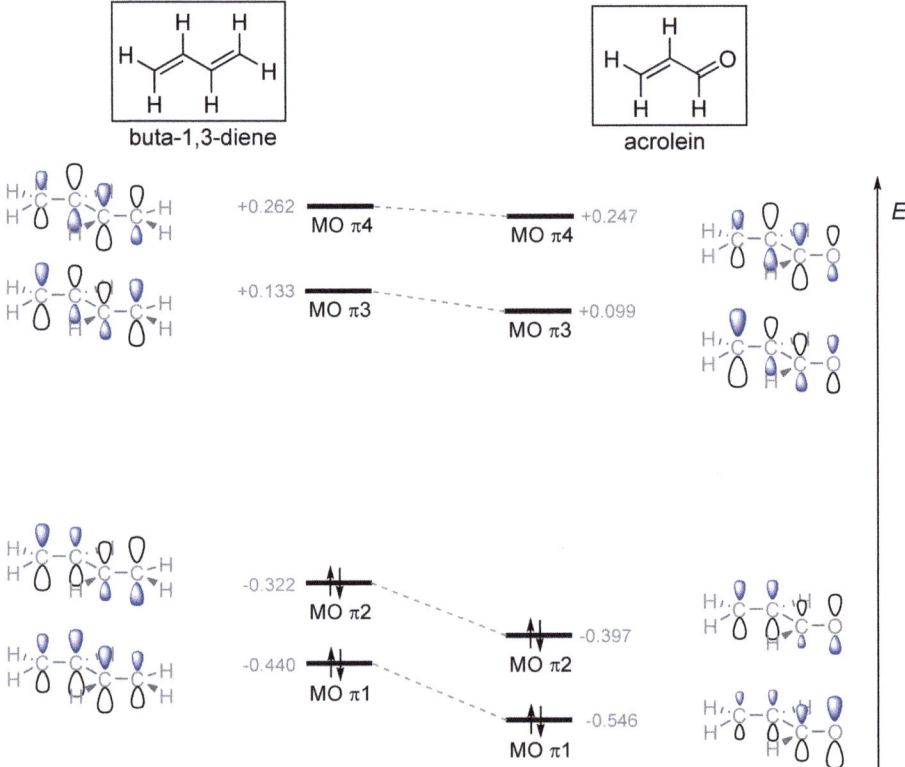

Figure 1.18 Molecular orbitals and orbital energies (in Hartree) for the π-systems in buta-1,3-diene (left) and acrolein (right).

represents the LUMO of the system. The highest lying $\pi 4$ orbital is dominated by the central two carbons and shows little contribution from the two outside centers.

In most organic molecules the molecular orbitals describing the spatial distribution of lone-pair electrons are located slightly above or just below the orbitals describing π-electrons. However, due to symmetry reasons the fragment orbitals describing lone-pair electrons combine with fragment orbitals of the σ-bonding framework such that overall delocalized molecular orbitals result. This can be nicely demonstrated for formaldehyde, whose five highest occupied orbitals are shown together with the LUMO in Figure 1.19. The calculated LUMO structure (that is, MO9) is quite similar to the schematic drawing of the $\pi 2$-orbital in Figure 1.16 and mainly located at the carbonyl carbon atom. The corresponding π-orbital composed of p-type atomic orbitals at carbon and oxygen is MO7 and thus the next-highest occupied molecular orbital

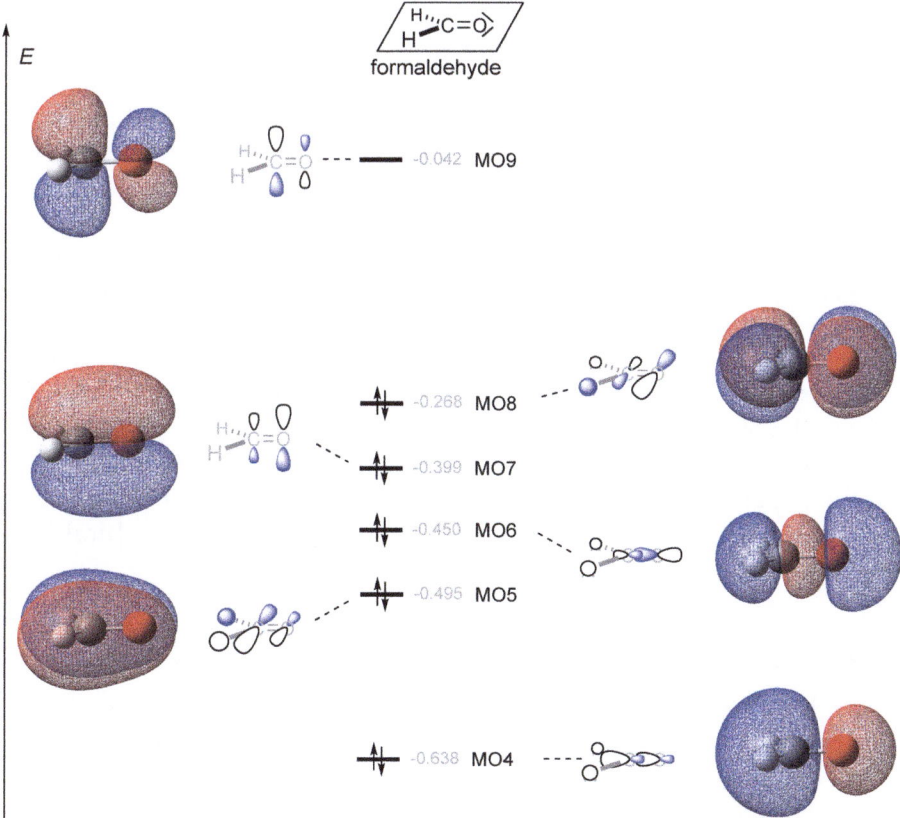

Figure 1.19 Molecular orbitals and orbital energies (in Hartree) for the π-system and the non-bonding electrons in formaldehyde as obtained at the B3LYP/6-31G(d) level of theory.

(NHOMO) of the overall system. The actual HOMO of the system (MO8) is delocalized in nature and constructed from a p-type atomic orbital on oxygen and an appropriately symmetric methylene group orbital. Despite its delocalized nature, MO8 has its main contribution from the oxygen atom and thus qualifies as a lone-pair orbital. The same two fragments also combine (with opposite sign) in the construction of MO5, a molecular orbital mainly describing the methylene group C–H bonds. This leaves us with MO6 and MO4 as another set of orbitals with delocalized structure, whose building blocks are best understood as another methylene group fragment orbital and an approximately sp²-type hybrid orbital on oxygen. These two fragments combine to yield a higher and a lower lying combination. The energetically higher lying MO6 orbital is mainly localized on oxygen and may be seen as the second lone-pair orbital of the system, while the lower lying MO4 describes mainly C–H bond formation. We can conclude from this survey that the molecular orbitals describing lone-pair electrons at carbonyl groups differ substantially in their shape and energy, and that combination with the adjacent σ framework occurs such that delocalized orbital structures result. On first sight this description of two spatially and energetically different lone-pair orbitals on oxygen is in conflict with the "rabbit ear" Lewis structures employed in qualitative valence bond theory, which is often assumed to imply energetic equivalence.

However, the two canonical MOs describing lone-pair electrons can be transformed without energy gain or loss into an alternative set of energetically equivalent MOs. It is thus not possible to exclude one or other of the descriptions of lone-pair orbitals without reference to a particular observable (or experiment). This experiment usually leads to a change in the bonding situation of the molecular system, and conclusions concerning the nature of the latter are therefore indirect (at best). Still, for the sake of consistency with molecular orbitals obtained from *ab initio* or density functional theory (DFT) calculations, in the following discussion we will present lone-pair orbitals in carbonyl compounds as shown pictorially in Figure 1.19.

1.3.4 Interaction of π- and σ-Bonded Systems

The rationalization of inductive substituent effects, such as those responsible for the stabilization of radicals or carbocations through attached alkyl groups, relies on the effective interaction between fragment orbitals of the alkyl group and molecular orbitals of the π-system. The nature of these interactions will, in the following discussion, be demonstrated using a methyl-substituted carbocation as an example. The methyl substituent attached to the cation center contains one carbon and three hydrogen atoms. Using the three H(1s), the C(2s) and the three C(2p) atomic orbitals as building blocks, we can construct the seven molecular orbitals of the methyl group using the principal symmetry plane (σ_v) of the carbocation as a construction aid (Figure 1.20). Molecular orbitals of the methyl fragment will thus be either symmetric (**S**) or antisymmetric (**A**) with respect to this symmetry plane. Furthermore, three of the seven fragment MOs show threefold symmetry around the x-axis of the Cartesian coordinate system and are therefore termed σ-orbitals. The remaining four fragment MOs are π-orbitals with different numbers of nodal planes. The lower two of these π-MOs are isoenergetic and fully occupied with two electrons each, while the upper two π-MOs are also isoenergetic, but empty. Between these two sets of π-orbitals we find the singly occupied σ-orbital through which the methyl group will attach to the molecular framework of the cation. When forming a σ-bond to the cation center the energy of methyl fragment MO σ2 will drop to similar levels to that of methyl fragment MO σ1. This leaves the occupied π-type orbitals π1 and π2 as the highest lying occupied MOs of the methyl group and thus the most important interaction partners with the vacant C($2p_z$) orbital at the cation center. This latter interaction is guided by the common symmetry elements of the respective (fragment) orbitals, as shown in Figure 1.21.

On the side of the methyl group the important players are the filled π-type orbitals π1 and π2. These fragment MOs differ in their symmetry properties in that π1 is symmetric (**S**) and π2 is antisymmetric (**A**)

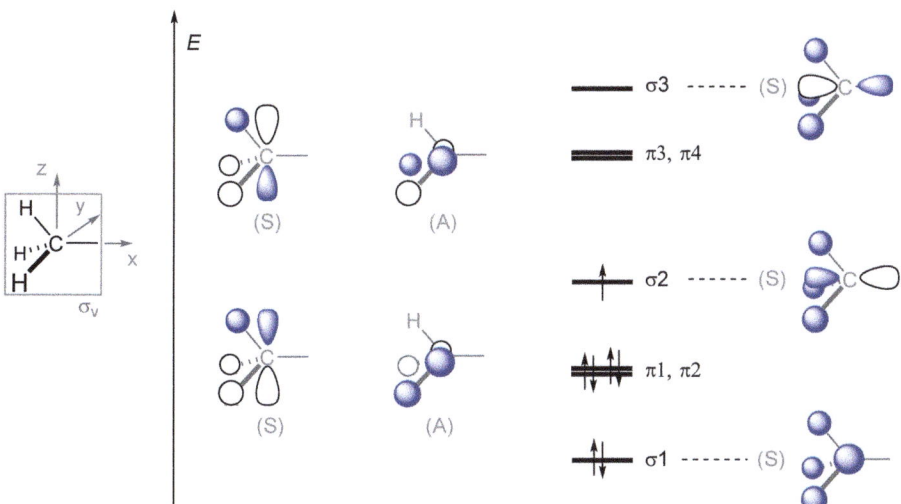

Figure 1.20　Molecular orbitals of the methyl group.

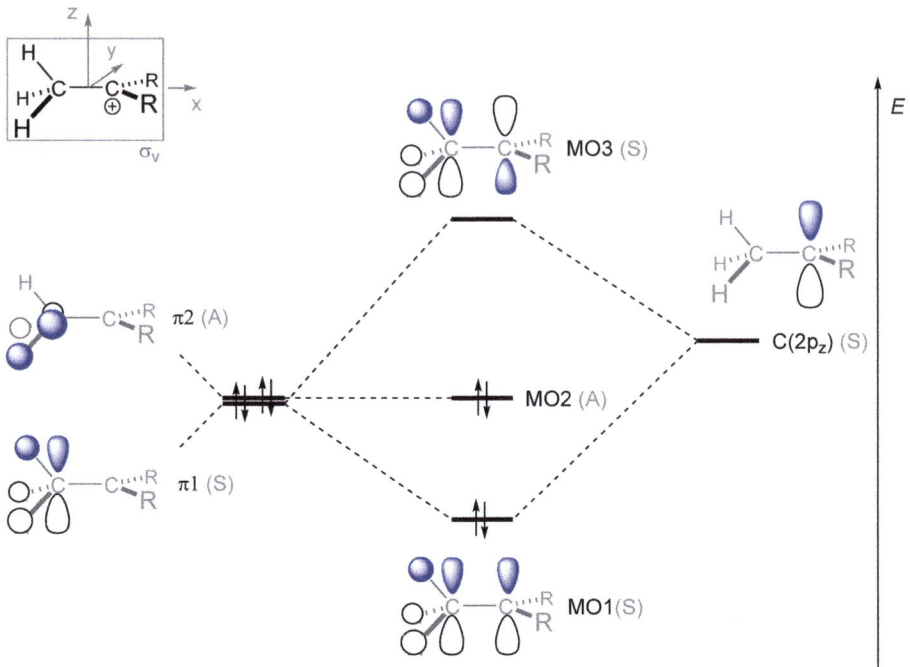

Figure 1.21 Interaction of methyl group orbitals with the cation center in carbocations.

with respect to the vertical symmetry plane (σ_v) of the carbocation. The vacant $C(2p_z)$ orbital available at the cation center is also symmetric with respect to the σ_v symmetry plane and combination with the methyl group MO $\pi 1$ thus generates two new molecular orbitals MO1 and MO3 (Figure 1.21). Both of these are symmetric with respect to the common symmetry σ_v plane and the lower lying MO1 is filled with two electrons. The antisymmetric methyl group fragment MO $\pi 2$ remains unchanged at this point as it cannot interact with the $C(2p_z)$ orbital at the cation center. What is shown in Figure 1.21 as the specific interaction of methyl group fragment orbitals with a carbocation center can easily be adapted to other situations, where alkyl groups attach to π-systems through simply replacing the $C(2p_z)$ orbital with the appropriate vacant π-MOs.

On a more general note we can summarize the valence space bonding situation in typical "organic" molecules mainly constructed from hydrogen and first-row elements, such as carbon, oxygen and nitrogen, in the following way: the lowest lying molecular orbitals are those describing σ-bonds between first-row elements (C–C, C–O, *etc.*), followed by σ-bonds involving hydrogen (C–H bonds). Higher orbital

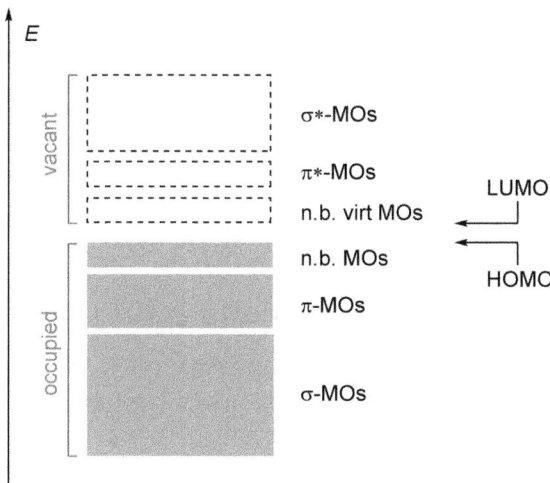

Figure 1.22   General scheme for orbital energies in organic molecules.

energies are typically found for molecular orbitals for π-electrons, followed by those for non-bonding (n.b.) electrons. The lowest vacant MOs (often also termed virtual MOs) are those deriving from the respective atomic orbitals lacking a bonding partner, as exemplified by the bonding situation at the carbocation center in Figure 1.21. At higher orbital energies we then find the π^*-MOs and at still higher energies the σ^*-MOs generated on bond formation through combination of two fragment or atomic orbitals into a bonding and an antibonding combination (Figure 1.22). The highest lying occupied molecular orbital is commonly referred to as the HOMO, and the lowest unoccupied molecular orbital as the LUMO. In discussion of molecular reactivity it is often helpful (and possibly also necessary) to inspect the properties of the NHOMO located directly below the HOMO. In a similar vein, the next orbital located above the LUMO (the NLUMO) may be interesting. In organic radicals the distribution of the unpaired electron will (mainly) be described by the SOMO. How all of these orbitals interact with those of a reaction partner along a particular reaction coordinate is the subject of frontier molecular orbital (FMO) theory (see Chapter 2, Section 2.7 for further discussion).

1.4 Hyperconjugation

In the language of molecular orbital theory, hyperconjugation results from orbital interactions beyond those of primary bond-making character. In the five cases discussed in the following sections (stabilization of cations and radicals, anomeric effect, Bohlmann bands, *gauche* effect, conformational barriers in alkanes) energetically favorable interactions result from the interaction of a filled orbital (the donor) with an empty orbital (the acceptor). In all these cases the donor/acceptor interactions require specific relative orientations of the participating orbitals, which leads to stereochemical consequences for the respective molecular systems.

1.4.1 Stabilization of Carbocations and Radicals

How carbocations interact with adjacent alkyl groups has already been discussed in Section 1.3.4 in terms of the participating fragment orbitals. Occupied C–H bond orbitals of π-symmetry act as donor orbitals in this case, while a vacant C(2p) orbital at the cation center is the acceptor. In the language of VB theory the same interactions can be described by combination of the canonical Lewis structure **C1** with the "no-bond" Lewis structure **C2** (Figure 1.23a). For carbon-centered radicals a completely analogous set of Lewis structures can be employed to rationalize their stabilization through alkyl group substituents. While this is not immediately obvious from a comparison of the two sets of Lewis structures in Figure 1.23a and b, alkyl group substituent effects are much larger in cations compared with radicals.

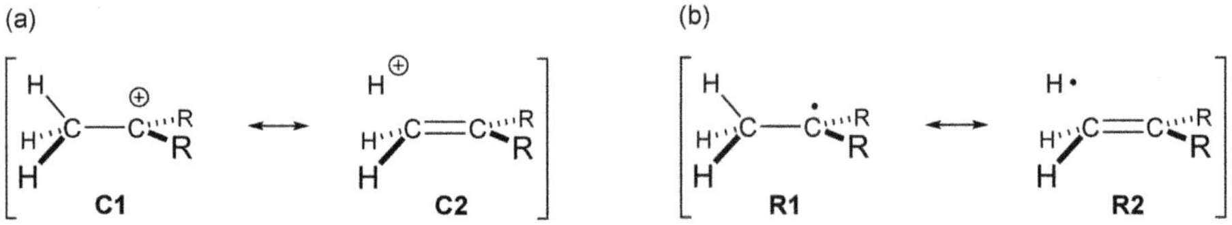

Figure 1.23 VB description of the stabilization of (a) carbocations and (b) carbon-centered radicals through hyperconjugative interactions.

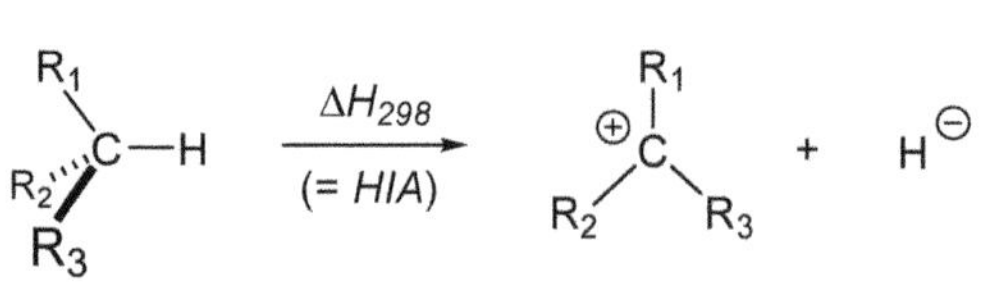

$$(1.1)$$

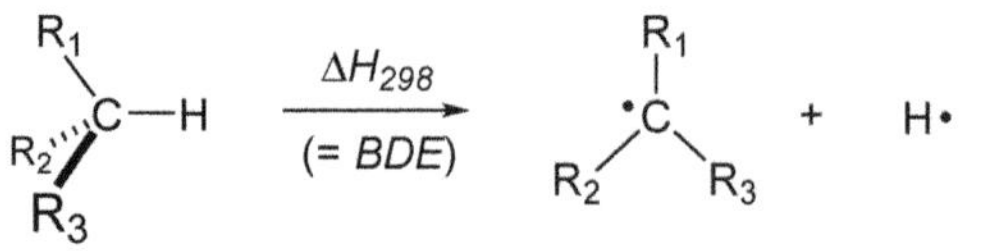

$$(1.2)$$

The stability of these transient species can be quantified through values for the heterolytic or homolytic C–H bond dissociation energies in the respective parent hydrocarbons. Heterolytic C–H bond dissociation is described by eqn (1.1), and the reaction enthalpies are commonly referred to as hydride ion affinities (HIA) of the respective carbocations. This formal C–H bond heterolysis is highly endothermic, with larger values indicating less stable carbocations, as shown in Figure 1.24 for a number of relevant systems. The reference against which all substituted systems is compared is the methyl cation with HIA = +1314 kJ mol⁻¹. Attachment of the first methyl group generates the ethyl cation with HIA = +1132 kJ mol⁻¹. Interaction between the methyl group substituent and the cation center is, in this case, substantial enough to lead to an overall bridged structure. Adding further methyl substituents, as in the isopropyl (HIA = +1054 kJ mol⁻¹) and *tert*-butyl cations (HIA = + 991 kJ mol⁻¹) leads to much more stable carbocations with classical (unbridged) structures. From the pictorial representation of these affinity values in Figure 1.24 it can easily be seen that the stabilizing effects of methyl substituents become increasingly smaller after each substitution step. This "saturation" effect can be rationalized with the orbital interaction scheme shown in Figure 1.21, where the donor/acceptor interaction of the substituent with the cation center leads to an increase in the energy of the vacant acceptor orbital. Donor/acceptor interactions for the next substituent attached to the cation center will thus be less effective due to the higher lying acceptor orbital. Stabilization of carbocations through attached π-systems, as in the allyl cation (HIA = +1071 kJ mol⁻¹) or the benzyl cation (HIA = +1002 kJ mol⁻¹) are larger than those for the simple alkyl groups. The cumulative effects of three alkyl substituents, however, still make the *tert*-butyl cation more stable than the allyl and benzyl cations.

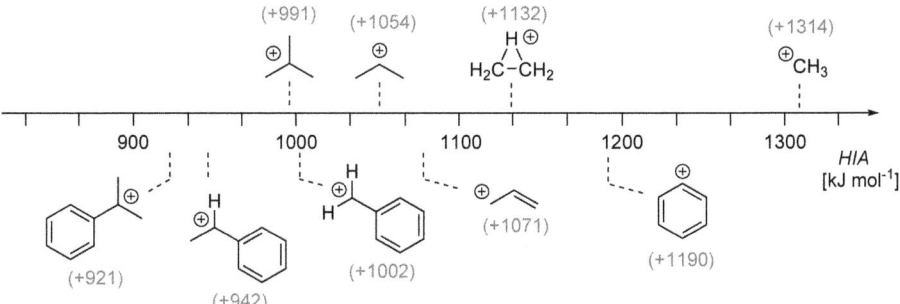

Figure 1.24 Hydride ion affinity (HIA) values for selected carbocations.[2,3]

Homolytic bond dissociation energies (BDE) for the same hydrocarbons as those used to generate the carbocations in Figure 1.24 are much smaller in absolute terms. The reference against which substituent effects will be compared is the C–H bond in methane, with a homolytic BDE(C–H) of +439 kJ mol^{-1} (Figure 1.25). The resulting methyl radical is a planar species that can interact with attached substituents through inductive, resonant or charge-transfer interactions. Addition of the first methyl substituent leads to the ethyl radical, which is characterized by a classical (non-bridged) structure and BDE = +420 kJ mol^{-1}. The stabilizing effect of the methyl group thus amounts to 19 kJ mol^{-1}, compared to 182 kJ mol^{-1} in the case of the ethyl/methyl cation. This substantial difference can readily be rationalized by the MO interaction diagram in Figure 1.21 for the methyl group/cation interaction. The same MO diagram also applies to the methyl group/radical case, the difference being that the participating C(2p) orbital now contains one electron. On combination with the methyl group orbitals and generation of the new bonding and antibonding MOs, the extra electron will be located in the antibonding MO3. Moving electron density to antibonding molecular orbitals is energetically costly and the reason for the very different substituent effects in cations and radicals. Addition of a second and a third methyl group, as in the isopropyl radical (BDE = +410 kJ mol^{-1}) and the *tert*-butyl radical (BDE = +400 kJ mol^{-1}), leads to more stable radicals, with the absolute as well as the relative magnitude of the substituent effects now being even smaller than before. Carbon-centered radicals are stabilized quite effectively by substituents containing π-systems, as is readily seen for the allyl radical (BDE = +369 kJ mol^{-1}) and the benzyl radical (BDE = +375 kJ mol^{-1}). In both cases the stabilizing effects

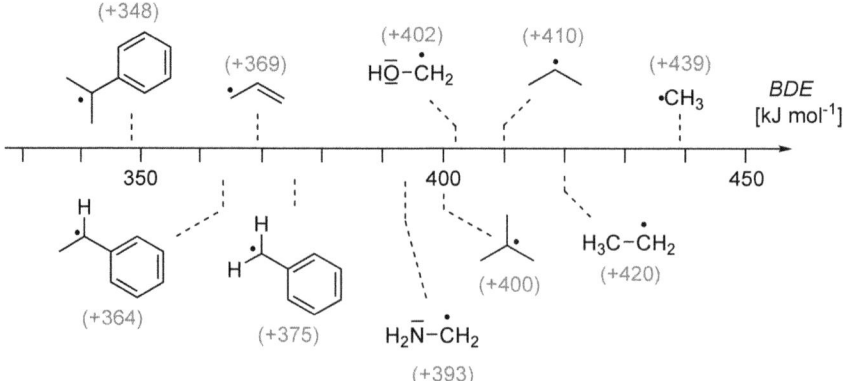

Figure 1.25 Homolytic BDE(C–H) data for selected radicals.[4]

derive from resonant delocalization of the unpaired spin. A third mechanism for radical stabilization appears in the C-centered radicals derived from methanol (BDE(C–H) = +402 kJ mol^{-1}) and methyl amine (BDE(C–H) = +393 kJ mol^{-1}). Here, the lone-pair-carrying heteroatoms donate some electron density to the radical center through charge-transfer interactions.

1.4.2 The Anomeric Effect

The anomeric effect was first described in the context of energy differences of carbohydrate isomers, where it was found that isomers with axial substituents at the C1 position (the anomeric center) are more stable than would be expected on the basis of steric effects alone. An often cited example concerns the 36 : 64 ratio between α- and β-D-glucopyranose in aqueous solution under equilibrating conditions (Figure 1.26a). At 298.15 K this ratio corresponds to a free energy difference in favor of the β-isomers of 1.6 kJ mol^{-1}. Whether or not this is remarkable depends on our expectations, and these are often defined by the "*A*-values" (or conformational free energy differences) for the respective substituents in cyclohexane ring systems. These can be determined using conformationally locked cyclohexanes, such as the 4-*tert*-butylcyclohexanols, as shown in Figure 1.26b. In hydroxylic solvents, such as water and simple alcohols, the axial orientation of the OH group is disfavored over the equatorial orientation by 4.0 kJ mol^{-1} for these systems. Taking this value as a

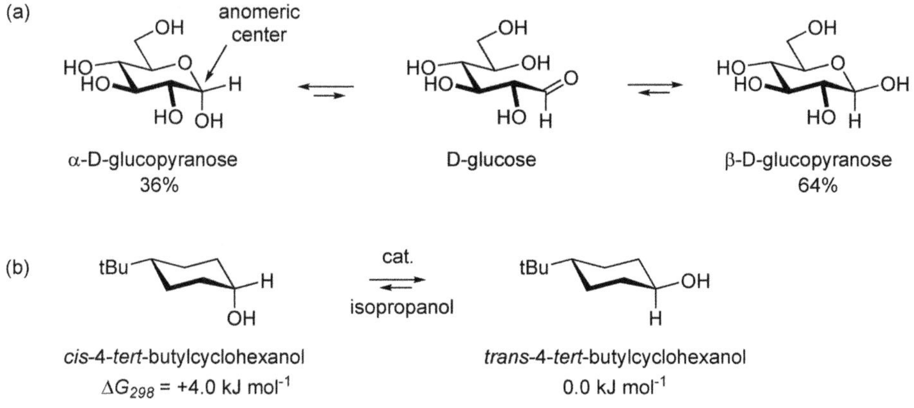

Figure 1.26 (a) Ratio between α- and β-D-glucopyranose in D$_2$O and (b) relative free energies ΔG_{298} for *cis* and *trans* isomers of 4-*tert*-butylcyclohexanol.[5-7]

reference, the anomeric effect in glucopyranoses thus amounts to $4.0 - 1.6 = 2.4$ kJ mol^{-1}.

The energy difference between constitutional isomers reappears in simpler model systems as the difference between conformational isomers. Again, reference is made to *A*-values of systems lacking stereoelectronic effects. For the methoxy group in methoxycyclohexane the *A*-value amounts to +3.1 kJ mol^{-1}, the positive sign indicating a preference for the equatorial form. The presence of a ring oxygen atom, as in 2-methoxytetrahydropyran, reverses this trend, now favoring the axial conformer by 3.8 kJ mol^{-1}. The anomeric effect can thus be quantified in this case as the difference between these values of 6.9 kJ mol^{-1} (Figure 1.27). Theoretical estimates for these systems at a highly correlated level put the energy component of the anomeric effect at 0 K (that is, excluding thermal and entropy effects) at 6.2 kJ mol^{-1}.

An intriguing additional phenomenon observed for 2-methoxytetrahydropyran, as well as for many other systems, is the solvent dependence of its conformational preference. While the equatorial:axial (eq:ax) ratio amounts to 18:82 in apolar solvents, such as benzene, a lower value of 29:71 is observed in CHCl$_3$, dropping further to 48:52 in polar solvents such as water. The high proportion of the axial conformer, as well as the influence of the medium, is rationalized by the combination of two effects. The first of these is electrostatic in nature and based on the significantly higher dipole moment of the equatorial conformer compared with the axial one. In the absence of a polar medium (in the gas phase) this leads to a preference for the axial conformer. The presence of a polar medium stabilizes the equatorial conformer more than the axial one and thus lowers the energy difference between both isomers. The second model is based on molecular orbital interactions involving the occupied lone-pair (or non-bonding) orbital of the ring oxygen atom and the σ^*(C–O) orbital of the exocyclic C–O bond (Figure 1.28).

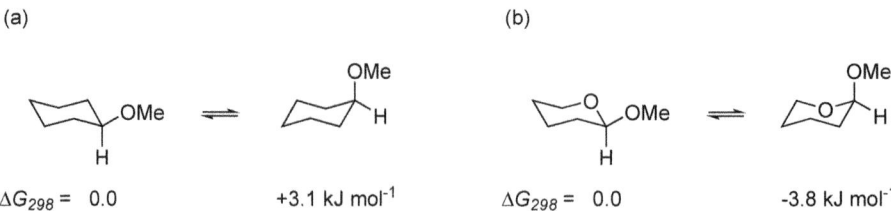

(a)   (b)

OMe   OMe

ΔG_{298} = 0.0   +3.1 kJ mol^{-1} ΔG_{298} = 0.0   -3.8 kJ mol^{-1}

Figure 1.27 Free energy differences between equatorial and axial conformers of (a) methoxycyclohexane and (b) 2-methoxytetrahydropyran.

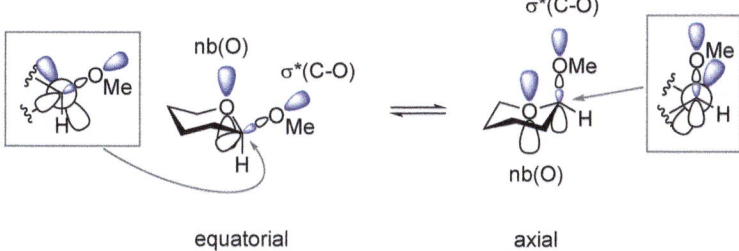

Figure 1.28 Donor/acceptor interactions between the occupied non-bonding orbitals at the ring oxygen atom (nb(O)) and the σ*(C–O) orbital of the exocyclic C–O bond in the axial and equatorial conformers of 2-methoxytetrahydropyran.

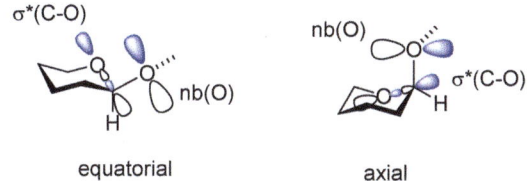

Figure 1.29 Donor/acceptor interactions between the occupied non-bonding orbitals at the exocyclic oxygen atom (nb(O)) and the σ*(C–O) orbital of the ring C–O bond in the axial and equatorial conformers of 2-methoxytetrahydropyran.

The alignment of these orbitals (and thus their overlap) is significantly better in the axial as opposed to the equatorial conformer, leading to more pronounced donor/acceptor interactions in the axial conformer as compared to the equatorial one.

Closely related to the anomeric effect (which is sometimes also referred to as the *endo*-anomeric effect) is the *exo*-anomeric effect in carbohydrates and related systems. This involves interaction between the oxygen lone-pair orbitals of the exocyclic oxygen atom and the σ*(C–O) orbitals of the ring C–O bond (Figure 1.29). This interaction is maximized in 2-methoxytetrahydropyran in conformations involving a *gauche* orientation between the ring C–O bond and the exocyclic O–CH₃ bond. As shown in Figure 1.29, the *exo*-anomeric effect can occur in both the axial and the equatorial conformers of this system and leads to defined conformational preferences for rotation around the exocyclic C–O bond. For this case, the size of the *exo*-anomeric effect can be estimated as the energy difference between the *gauche* conformers shown in Figure 1.29 and the corresponding *anti* conformers. This energy difference amounts to

17.2 kJ mol^{-1} for the axial conformer and 18.8 kJ mol^{-1} for the equatorial conformer at the MP2(FC)/6-311G(2df,2pd) level of theory. Analysis of a larger number of solid-state structures shows that the (*endo*-)anomeric and the *exo*-anomeric effects have a profound influence on the conformational properties of carbohydrates around the C1 center.

The anomeric effect is also responsible for the thermochemical stability of molecules carrying more than one electronegative substituent at the same carbon atom. Examples that are often cited are those of dimethoxymethane and difluoromethane, both of which can (formally) be constructed from their respective monosubstituted methane derivatives through a group exchange reaction (Figure 1.30). For dimethoxymethane this involves the reaction of two dimethylether molecules to generate dimethoxymethane, along with one molecule of methane. The number of bonds of equal type in this "isodesmic reaction" is identical on both sides of the equation (12 × C–H and 4 × C–O), which minimizes impact of methodological artifacts and inaccuracies on the final result. The overall heat of reaction of −54.9 kJ mol^{-1} is rationalized in terms of the same donor/acceptor interactions between the nb(O) and the σ*(C–O) orbitals already described in Figure 1.28 and 1.29. These are present in the dimethoxymethane product, but absent in the dimethylether reactant molecules. The reaction enthalpy of ΔH_{298} = −58.3 kJ mol^{-1} for the fluorine exchange reaction between two fluoromethane molecules to give difluoromethane and methane can be rationalized in a completely analogous way. In this case the donor/acceptor interactions involve the nb(F) and the σ*(C–F) orbitals in the difluoromethane product (Figure 1.30b). This type of analysis can be taken further to quantify the multiple donor/acceptor interactions between the three C–F bonds in trifluoromethane through an isodesmic reaction involving fluorine exchange between three fluoromethane reactants, as ΔH_{298} = −141.7 kJ mol^{-1} (Figure 1.30c).

The examples in Figure 1.30 already indicate that the anomeric effect extends far beyond the realm of carbohydrate chemistry, and a **"generalized anomeric effect"** can consequently be defined involving interactions between a non-bonding electron-pair donor orbital nb(X) and an appropriately placed and oriented σ*(A–Y) acceptor orbital. The corresponding orbital interaction diagram is shown in Figure 1.31. The magnitude of the generalized anomeric effect will depend on the energy gap between the participating donor and acceptor orbitals, with a smaller gap translating into higher interaction energies. Particularly effective donor substituents are those carrying energetically high lying non-bonding electron pairs, such as formally negatively

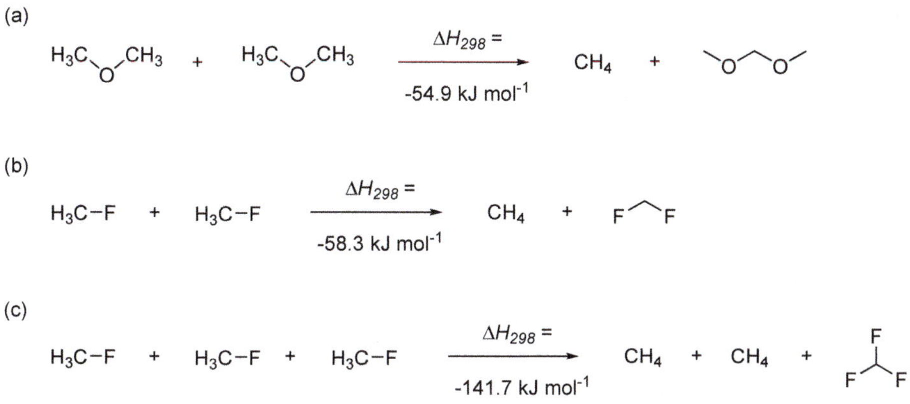

(a)

$$H_3C \diagdown O \diagup CH_3 \quad + \quad H_3C \diagdown O \diagup CH_3 \quad \xrightarrow[\text{-54.9 kJ mol}^{-1}]{\Delta H_{298} =} \quad CH_4 \quad + \quad \diagup O \diagdown \diagup O \diagdown$$

(b)

$$H_3C-F \quad + \quad H_3C-F \quad \xrightarrow[\text{-58.3 kJ mol}^{-1}]{\Delta H_{298} =} \quad CH_4 \quad + \quad F \diagup \diagdown F$$

(c)

$$H_3C-F \quad + \quad H_3C-F \quad + \quad H_3C-F \quad \xrightarrow[\text{-141.7 kJ mol}^{-1}]{\Delta H_{298} =} \quad CH_4 \quad + \quad CH_4 \quad + \quad F \diagdown C F_3$$

Figure 1.30 Reaction enthalpies for the isodesmic reactions generating (a) dimethoxymethane, (b) difluoromethane and (c) trifluoromethane.[1,8]

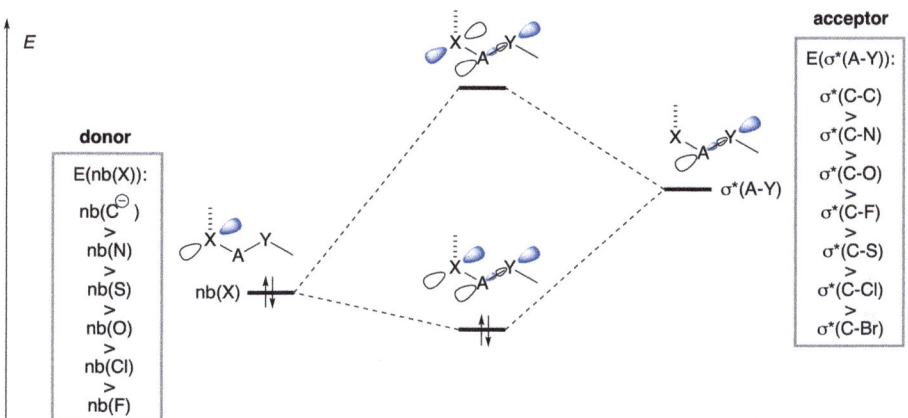

Figure 1.31 Orbital interaction diagram for the generalized anomeric effect.

charged carbanion centers or amines. In comparison, non-bonding electron pairs located at more electronegative elements, such as oxygen or fluorine, are less effective.

On the acceptor side the lowest lying σ*(C–Y) orbitals are those for the halides Y = Br and Cl, followed by S as another second-row element. The least efficient (highest lying) σ*(C–Y) orbitals are those for C–C bonds. The donor and acceptor components described in Figure 1.31 can be combined freely in the sense of an "anomeric effect toolkit" to predict the properties of new systems.

1.4.3 Bohlmann Bands

The appearance of unusually low C–H stretching vibrations in the infrared (IR) spectrum of aliphatic amines has been attributed, by F. Bohlmann, to donor/acceptor interactions of non-bonding electron pairs at nitrogen with suitably aligned neighboring σ*(C–H) orbitals (Figure 1.32). This interaction weakens the C–H bond and shifts the corresponding vibrational stretching frequency by 50–100 cm^{-1} to lower wavenumbers. These "Bohlmann bands" are thus located in a normally vacant part of the vibrational frequency spectrum (below 2800 cm^{-1}) and are characteristic for C–H bonds oriented in an *anti*-periplanar fashion to nitrogen lone-pair orbitals. A structural consequence of the Bohlmann effect is a moderate lengthening of the corresponding C–H bonds, by about 5%.

1.4.4 The *Gauche* Effect

The tendency of vicinal polar bonds to preferentially adopt a *gauche* conformation is called the "***gauche* effect**". An often cited example is 1,2-difluoroethane, whose *gauche* conformation is enthalpically preferred over the *anti* conformation by 3.2 kJ mol^{-1} in the gas phase. That this is remarkable becomes apparent on comparison with *n*-butane, whose *gauche* conformation is less stable than the *anti* conformation by 3.8 kJ mol^{-1} (even though this comparison is somewhat

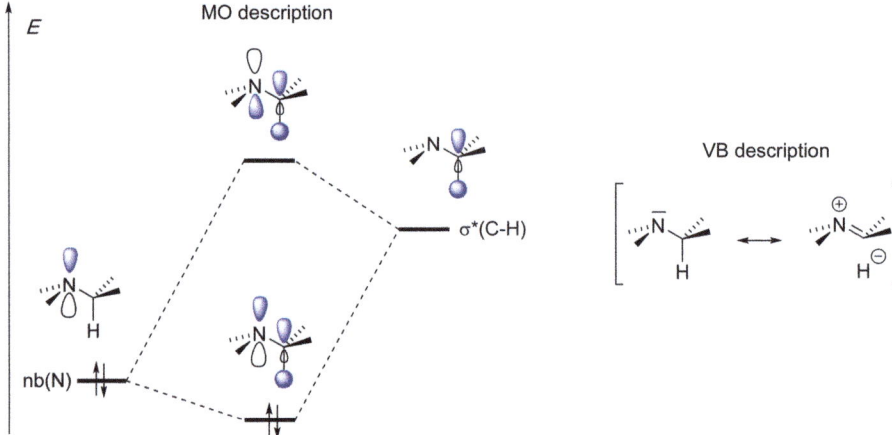

Figure 1.32 MO and VB descriptions for the donor/acceptor interactions responsible for the appearance of Bohlmann bands in the IR spectrum of nitrogen-containing natural products.

arbitrary). This effect can, to a large extent, be explained by the favorable donor/acceptor interactions of vicinal (neighboring) σ(C–H) and σ*(C–X) bonds. In the *gauche* conformation two such interactions are possible, while little can be done with the σ*-acceptor power of C–F bonds in the *anti* conformation. Alternatively, the same interactions can be described by the combination of the two Lewis structures in Figure 1.33. Solvent effects will enhance the population of the *gauche* conformation even more, since the solvation free energy of the *gauche* conformer is larger than that of the *trans* conformer. This is due to the higher symmetry of the *trans* conformer, leading to a zero dipole moment and consequently to a lower solvation energy. In most other systems containing vicinal polar bonds (such as C–F, C–O, C–Cl, C–N, C–Br or even C–CN, *etc.*) the weight of the *gauche* conformer is not as high as in 1,2-difluoroethane and the *gauche* conformer may not even be more stable than the *anti* conformer. Still, the orbital interactions described in Figure 1.33 are present in all of these systems, making the *gauche* conformer more stable than in the absence of these effects. Comparison is usually made to *n*-butane as the "non-interacting" reference system.

A preference for *gauche*-type conformations is also observed in systems where two vicinal polar bonds are joined together at their more negative ends. One of the best-studied systems is hydrogen peroxide, where the preferred (gas phase) conformation is characterized by a H/O/O/H dihedral angle of 112°. The corresponding *trans* conformation is less stable by 4.6 kJ mol^{-1} and a rotational transition state, while the *syn* conformation is less stable by 30.7 kJ mol^{-1} and is also a transition state for rotation around the O–O bond (Figure 1.34).

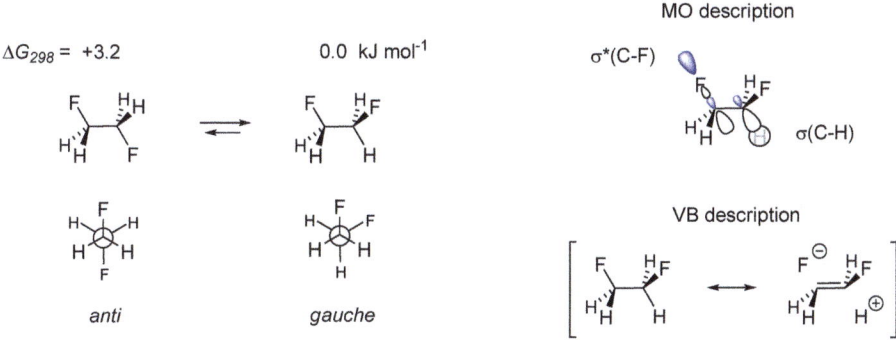

Figure 1.33 The *gauche* preference in 1,2-difluoroethane together with VB and MO descriptions of the principal donor/acceptor interactions.[9]

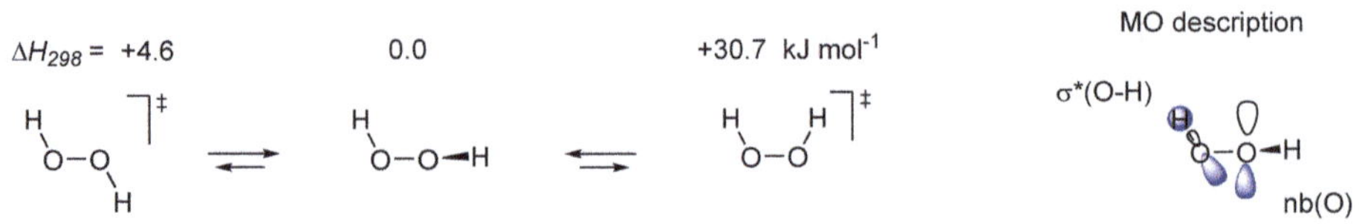

$\Delta H_{298} = $ +4.6   0.0   +30.7 kJ mol^{-1}

MO description

σ^*(O-H)

nb(O)

Figure 1.34 Conformational energies for hydrogen peroxide and the MO description of the principal donor/acceptor interactions.[10]

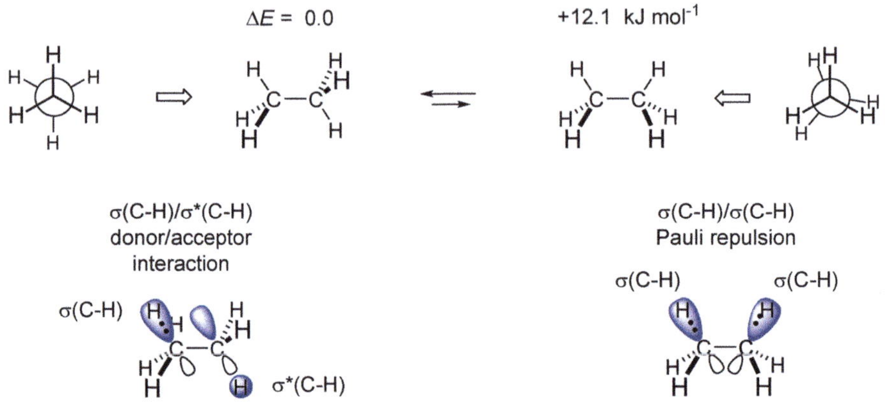

ΔE = 0.0   +12.1 kJ mol^{-1}

σ(C-H)/σ^*(C-H)
donor/acceptor
interaction

σ(C-H)/σ(C-H)
Pauli repulsion

σ(C-H) σ(C-H)

σ(C-H)

σ^*(C-H)

Figure 1.35 Conformational energies for ethane and principal orbital interactions responsible for conformational energies.[11]

What makes the *gauche*, or better "*skew*", conformation particularly stable is the favorable interaction of the σ^*(O-H) orbitals with the p-type lone-pair orbitals of the adjacent oxygen atom. This interaction is shut down in the *trans* conformer due to the orthogonal orientation of these orbitals, but maximizes its strength at dihedral angles around 90°. Analogous interactions exist in other systems of general formula A–B–C–D, with B/C being lone-pair bearing, electronegative elements, such as O, N or S, and A/D being electropositive elements, such as H or C. The preference for *skew*-type conformations in disulfides (RS–SR) and hydrazines (R$_2$N–NR$_2$) is also due to this effect.

1.4.5 Conformational Barriers in Alkanes

Conformational preferences in alkanes are currently discussed as an interplay of repulsive interactions between filled orbitals (Pauli repulsion) and attractive donor/acceptor interactions. The simplest system for the discussion of these conformational effects is ethane, whose preferred staggered conformation is located 12.1 kJ mol^{-1} below the respective eclipsed conformation (Figure 1.35). The staggered conformation is stabilized by donor/acceptor interactions between (occupied) $\sigma(C–H)$ and (vacant) $\sigma^*(C–H)$ orbitals, whose interaction is maximized in the *anti*-periplanar orientation. At the same time the eclipsed conformation is destabilized by repulsive interactions between occupied $\sigma(C–H)$ orbitals. How much these individual interactions actually contribute to the overall rotational barrier is still being discussed and depends on the underlying methodology used for the quantification of individual orbital interactions.

1.5   The Stability of Organic Molecules

The definition of the stability of organic molecules can never be absolute, but requires a reference relative to which the stability is defined. One commonly used strategy employs the stability of the respective elements in their standard states as the zero point of energy. The stability of an organic molecule of general formula $C_aH_bO_cN_d$ (its "heat of formation" ΔH_f°) is then nothing else but the reaction enthalpy for the defining eqn (1.3), where the target molecule is assembled from its elements. The term "standard states" refers, in this case, to a reaction temperature of 298.15 K, carbon as (solid) graphite and hydrogen, oxygen and nitrogen as gases:

$$aC(\text{graph.}) + b/2H_2(1\text{atm}) + c/2O_2(1\text{atm}) + d/2N_2(1\text{atm}) \xrightarrow{\Delta H_f^\circ} C_aH_bO_cN_d$$

$$(1.3)$$

$$C(\text{graph.}) + 2H_2(1\text{atm}) \xrightarrow{\Delta H_f^\circ} CH_4 \qquad (1.4)$$

For the simple example of methane shown in eqn (1.4), the calculation of the (gas phase) heat of formation requires one carbon atom and two equivalents of H_2 in their respective standard states, and yields a value of $\Delta H_f^\circ(CH_4) = -74.4$ kJ mol^{-1}. When comparing thermochemical data for a larger number of organic molecules it becomes apparent that their respective heats of formation are directly linked to the number of functional groups and characteristic fragments. This can easily be demonstrated using the ΔH_f° values for linear alkanes shown in Figure 1.36. The "functional group" added repeatedly to grow ethane $(\Delta H_f^\circ = -83.8$ kJ mol$^{-1})$ to *n*-nonane $(\Delta H_f^\circ = -228.2$ kJ mol$^{-1})$ is the methylene unit $(-CH_2-)$. Each addition of this unit appears to be linked to an increase in the heat of formation of around 21 kJ mol^{-1}. This observation, together with the analysis of a larger body of experimental data, has led to the development of Benson's group increment

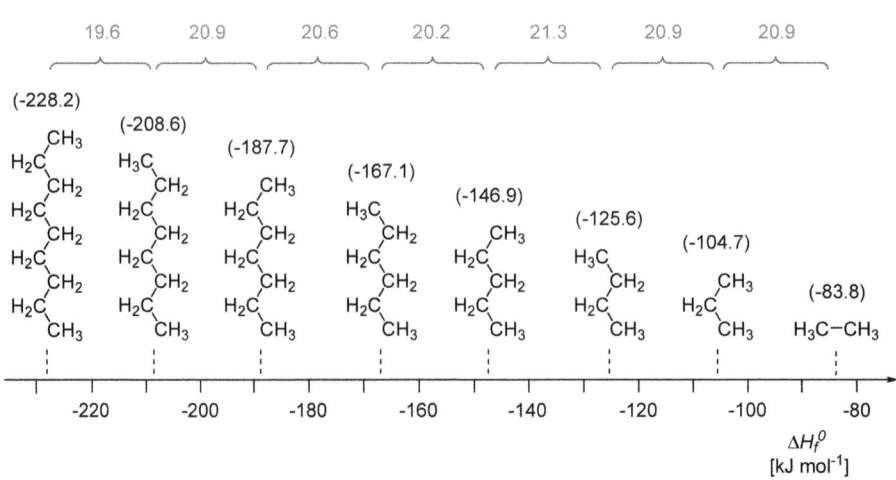

Figure 1.36   Heats of formation of selected linear alkanes (in black) together with the respective differences (in grey) between alkanes differing by one carbon atom (data from ref. 12).

Table 1.1   Group additivity values (GAVs, in kJ mol^{-1}) for fragments commonly found in organic molecules (data from ref. 12).

Group	ΔH_f° (kJ mol^{-1})	Group	ΔH_f° (kJ mol^{-1})
C-(C)(H)$_3$	−41.8	C-(C$_t$)(H)$_3$	−41.8
C-(C)$_2$(H)$_2$	−20.9	C$_B$–(H)	+13.8
C-(C)$_3$(H)	−10.0	C$_B$–(C)	+23.0
C–(C)$_4$	−0.4	C$_d$–(H)$_2$	+26.4
C-(C$_B$)(H)$_3$	−41.8	C$_d$-(C)(H)	+36.0
C-(C$_B$)(C)(H)$_2$	−19.2	C$_d$–(C)$_2$	+42.7
C-(C$_d$)(H)$_3$	−41.8	C$_t$–(H)	+113.8
C-(C$_d$)(C)(H)$_2$	−20.1	C$_t$–(C)	+114.2

method. According to this important concept the overall heat of formation of organic molecules can be calculated by simply summing up the group additivity values (GAVs) for all constituent functional groups. A list of common functional groups and their respective group increments is summarized in Table 1.1. The shorthand notation for functional groups X-(A)$_i$(B)$_j$(C)$_k$(D)$_l$ is based on a central atom X with a maximum valency of four, attached to i atoms of type A, j atoms of type B, *etc.* The "type" of an atom is chosen such that the additivity assumption underlying the group increment method holds with as few additional corrections as possible. For the hydrocarbons most frequently encountered in organic chemistry this leads to four types of carbon: C (sp^3-hybridized as in common aliphatic hydrocarbons),

C_d (sp^2 hybridized as in alkenes), C_t (sp-hybridized as in alkynes) and C_B (sp^2 hybridized as in benzenoid hydrocarbons). The group increment is given for each of these carbon types, together with the directly attached atom types. The methylene group mentioned before is based on a "C" central atom, connected to two additional carbon atoms of type "C" and two hydrogen atoms, and the proper description of this fragment is thus "C-(C)$_2$(H)$_2$". The connection list is shorter for carbon atoms in double or triple bonds simply due to the fact that one of the bonding partners is predefined through the respective multiple bond. The carbon atom at the end of a triple bond is thus described by type "C$_{t-}$(H)". Surveying the GAV values for different carbon types in Table 1.1, we note that these values are negative for groups with sp^3-hybridized carbon atoms, while they are positive for groups with sp^2- or sp-hybridized carbon atoms.

As an example of the application of Benson's GAV scheme we can use the increments listed in Table 1.1 to estimate the heat of formation of *n*-butane. This system contains two methylene groups and two terminal methyl substituents, and calculating the overall heat of formation using the increments in Table 1.1 thus yields $2 \times$ C-(C)$_2$(H)$_2$ $+ 2 \times$ C-(C)(H)$_3$ = $2 \times -20.9 + 2 \times -41.8 = -125.4$ kJ mol^{-1}. The experimentally determined value equates to ΔH_f°(*n*-butane) = -125.6 kJ mol^{-1}, which is in practically quantitative agreement with the prediction. Applying the same procedure to isobutane as a constitutional isomer of butane requires the C-(C)(H)$_3$ and C-(C)$_3$(H) increments from Table 1.1 and yields a predicted value of ΔH_f°(isobutane) = -135.4 kJ mol^{-1}, which is again closely similar to the experimental value of -134.2 kJ mol^{-1}. A similarly good agreement between calculated and experimentally measured values is also observed for toluene, hex-1-ene and propyne, as prototypical examples for benzenoid hydrocarbons, alkenes and alkynes (Figure 1.37). The very impressive performance of the GAV method observed here may, in part, be due to the fact that the GAV values have been derived from known thermochemical data of small hydrocarbons, that is, exactly the type of systems shown in Figure 1.37.

By design, the Benson GAV concept is flexible enough to also encode the more sophisticated 1,2- and 1,3-interactions present in molecular systems. The higher stability of isobutane compared with *n*-butane, for example, is thought to derive from a stabilizing 1,3-interaction between C(sp^3) carbon atoms (sometimes referred to as "protobranching"). There are three such interactions in isobutane, but only two in *n*-butane, hence the lower stability of the latter compared with the former.

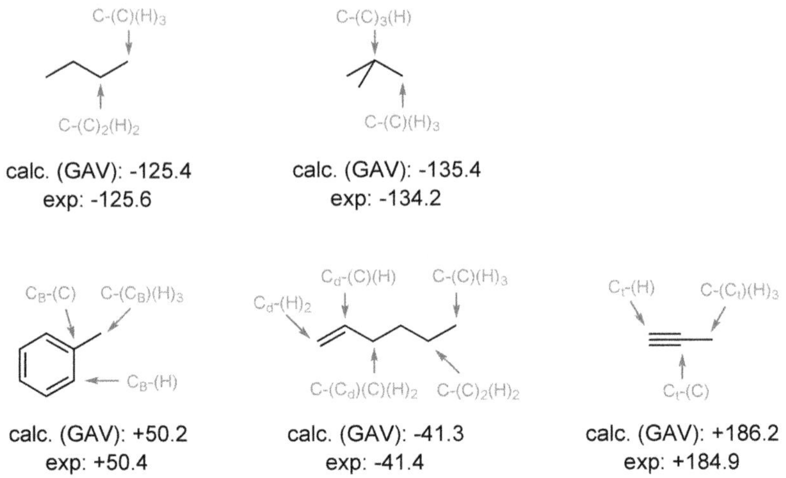

C-(C)(H)$_3$

C-(C)$_2$(H)$_2$

calc. (GAV): -125.4
exp: -125.6

C-(C)$_3$(H)

C-(C)(H)$_3$

calc. (GAV): -135.4
exp: -134.2

C$_B$-(C)   C-(C$_B$)(H)$_3$

C$_B$-(H)

calc. (GAV): +50.2
exp: +50.4

C$_d$-(C)(H)   C-(C)(H)$_3$

C$_d$-(H)$_2$

C-(C$_d$)(C)(H)$_2$ C-(C)$_2$(H)$_2$

calc. (GAV): -41.3
exp: -41.4

C$_t$-(H)   C-(C$_t$)(H)$_3$

C$_t$-(C)

calc. (GAV): +186.2
exp: +184.9

Figure 1.37 Heat of formation (ΔH_f°) values for selected hydrocarbons calculated with the GAV increments in Table 1.1 (data from ref. 12).

1.6 Strained Molecules

Benson's group increment method provides us with a systematic numerical procedure for predicting the heat of formation of organic molecules under, in principle, strain-free conditions. Any type of unfavorable steric interaction or strain within the molecule will lead to an increase in its heat of formation. This, in turn, provides us with a methodology for quantifying the degree of strain present in a given bonding situation. In the following discussion we will concentrate on the two most common types of strain in organic systems: strain caused through unfavorable 1,4-interactions and ring strain.

1.6.1 Unfavorable 1,4-Interactions

The heat of formation of 2-methylbutane can easily be calculated using the group increments listed in Table 1.1 as ΔH_f°(2-methylbutane) = 3 × (−41.8) − 20.9 − 10.0 = −156.3 kJ mol^{-1}. This value is too negative when compared to the experimental value of −153.7 kJ mol^{-1}. The deviation may be due to the presence of at least one unfavorable *gauche* interaction between the terminal methyl substituents, as is easily seen in a Newman projection along the C2–C3 bond axis. Addition of Benson's *gauche* correction of +3.3 kJ mol^{-1} to the heat of formation based on GAV increments alone predicts a new heat of formation of −153.0 kJ mol^{-1}, which is significantly closer to the experimental value than before (Figure 1.38a).

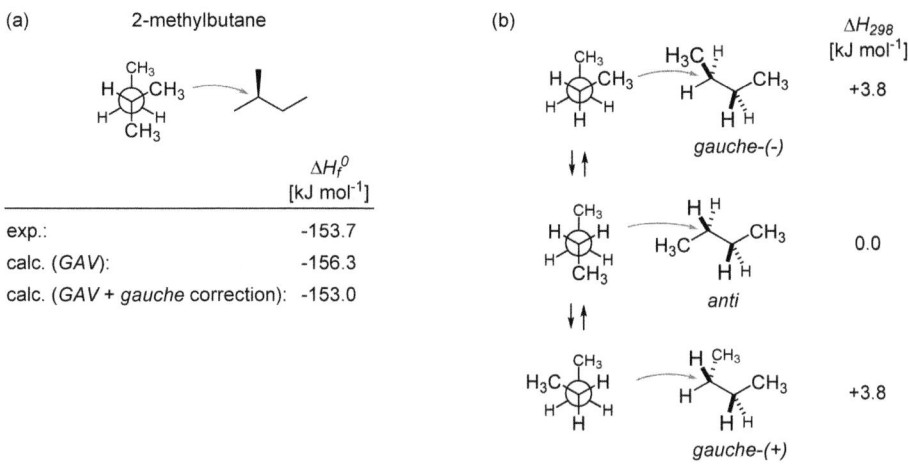

Figure 1.38 (a) Heat of formation of 2-methylbutane calculated without and with a correction for *gauche* interactions (data from ref. 12) and (b) enthalpy differences between *anti* and *gauche* conformations in *n*-butane.

Closely related to Benson's *gauche* correction is the *anti*/*gauche* conformational energy difference in linear alkanes. For *n*-butane, as the smallest alkane showing this effect, two *gauche* conformations exist of identical energy. In one of these, termed the "*gauche*-(−)" (or g−) conformation the C/C/C/C dihedral angle along the carbon backbone is around −60°, while a positive dihedral angle of identical magnitude is present in the "*gauche*-(+)" (or g+) conformation. Both *gauche* conformations are located 3.8 kJ mol^{-1} above the global minimum *anti* conformation (Figure 1.38b).

The stability difference between alkenes with (E) and (Z) configurations can obviously not be predicted using the GAV data in Table 1.1 alone, as the constituent functional groups are identical in both cases. Still, alkenes with (Z) configuration are typically less stable than those of (E) type, again a reflection of unfavorable 1,4-interactions. A case in point are the (E)- and (Z)-isomers of hex-3-ene, with experimentally measured heats of formation of ΔH_f°((E)-hex-3-ene) = −54.4 kJ mol^{-1} and ΔH_f°((Z)-hex-3-ene) = −47.6 kJ mol^{-1}. Using the increment data from Table 1.1 we predict a heat of formation of −51.8 kJ mol^{-1} for both isomers. The recommended Benson *cis* correction for alkenes amounts to +4.6 kJ mol^{-1}, and addition of this correction for the (Z)-isomer indeed leads to an improved prediction of ΔH_f°((Z)-hex-3-ene) = −47.2 kJ mol^{-1}. The correction term is, however, not quite as large as the stability difference determined experimentally, which also illustrates the limits of using a single correction term for all types of alkenes with a (Z) configuration (Figure 1.39).

The (E)/(Z) stability difference recurs in benzene derivatives (and similar aromatic systems) as the energy difference between *ortho*- and *meta*-isomers. For example, the stability difference between *ortho*-dimethylbenzene (1,2-dimethylbenzene or *ortho*-xylene, $\Delta H_f^{\circ} = +19.1$ kJ mol^{-1}) and its *meta*-isomer (1,3-dimethylbenzene or *meta*-xylene, $\Delta H_f^{\circ} = +17.3$ kJ mol^{-1}) is 1.8 kJ mol^{-1} when using experimental data. Heats of formation predicted with the GAV data in Table 1.1 alone are identical for both systems and correspond to $4 \times 13.8 + 2 \times 23.0 - 2 \times 41.8 = +17.6$ kJ mol^{-1}. This is quite close to the experimental value for the *meta*-isomer, but too low for the *ortho*-compound. Addition of the Benson *ortho* correction of +2.5 kJ mol^{-1} for 6- and 5-membered aromatic systems improves the prediction for the *ortho* compound system significantly, now giving a value of +20.1 kJ mol^{-1} (Figure 1.40).

Several non-nearest neighbor interactions exist beyond the 1–4 interactions described above, but their systematic incorporation into the Benson increment scheme becomes more difficult. One of the more relevant repulsive interactions in larger alkanes is the *syn*-pentane or

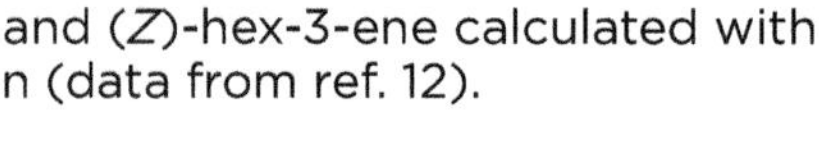

		(E)-hex-3-ene	(Z)-hex-3-ene
ΔH_f^0 [kJ mol^{-1}]	exp.:	-54.4	-47.6
	calc. (GAV):	-51.8	-51.8
	calc. (GAV + cis correction):	-51.8	-47.2

Figure 1.39 Heats of formation for (E)- and (Z)-hex-3-ene calculated without and with a *cis* correction (data from ref. 12).

		1,2-dimethylbenzene (ortho-dimethylbenzene)	1,3-dimethylbenzene (meta-dimethylbenzene)
ΔH_f^0 [kJ mol^{-1}]	exp.:	+19.1	+17.3
	calc. (GAV):	+17.6	+17.6
	calc. (GAV + ortho correction):	+20.1	+17.6

Figure 1.40 Heats of formation for *ortho*- and *meta*-dimethylbenzene calculated without and with an *ortho* correction (data from ref. 12).

"*g*+*g*−" interaction occurring in *n*-pentane. The four conformational minima present in this alkane (neglecting mirror images) are shown in Figure 1.41 mapped onto a cyclohexane grid for ease of comparison. In the most stable "aa" conformer both terminal methyl groups occupy an *anti* position relative to the central C3 fragment. Flipping one of these methyl groups into a *gauche* conformation leads to the second most stable conformer labeled "*g*+a", which is located 2.6 kJ mol^{-1} higher in energy. The third best conformer at +4.1 kJ mol^{-1}, labeled "*g*+*g*+" flips both terminal methyl groups into a *gauche* conformation such that they end up on opposite sides of the plane defined by the central C3 unit. The "*g*+*g*−" conformer, in contrast, positions both methyl groups next to each other on the same side of the central C3 unit and does not even represent a local minimum on the potential energy surface. Instead, one of the *gauche* dihedral angles opens up from around 60° to over 90° in order to relieve steric interactions with its neighbor. This "x+*g*−" conformation is located 11.4 kJ mol^{-1} above the global minimum and thus represents the energetically least favorable *n*-pentane conformer. Avoiding *g*+*g*− conformations is thus an important principle in the conformational control of otherwise flexible organic molecules.

Closely related to the *syn*-pentane interaction is the 1,3-allylic (or A^{1,3}) strain in alkenes. This effect was discussed by Johnson and Malhotra in 1965 following the inspection of molecular models for substituted methylenecyclohexanes. In the presence of a sterically demanding substituent, R, in the 2 position of the cyclohexane ring, and a second substituent, R′, occupying the (*Z*)-position at the double bond terminus, as shown in Figure 1.42a, the steric effects between these two substituents will be substantial in conformation **A2**, where substituent R occupies the equatorial position. Ring inversion to conformation **A1** rotates substituent R into an axial position and thus away from R′. Quantitative estimates for allylic strain effects can be taken from the conformational analysis of 3-methylbut-1-ene and

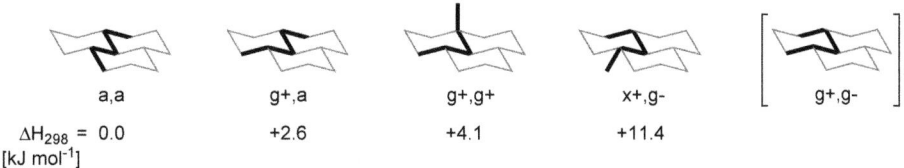

a,a	g+,a	g+,g+	x+,g-	[g+,g-]

ΔH_{298} = 0.0   +2.6   +4.1   +11.4
[kJ mol^{-1}]

Figure 1.41 Conformational minima of *n*-pentane (shown in black) mapped onto a cyclohexane grid (shown in grey) together with conformational energy differences (data from ref. 13).

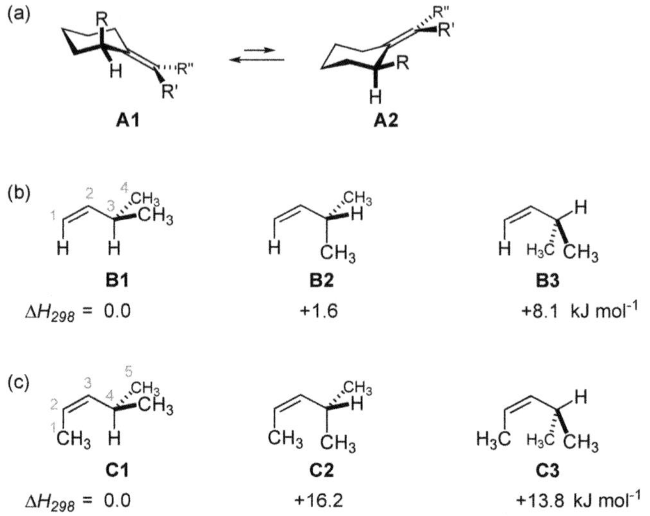

Figure 1.42 Conformational properties of (a) substituted methylenecyclohexanes, (b) 3-methylbut-1-ene and (c) (*Z*)-4-methylpent-2-ene. (Conformational energies in (b) and (c) have been calculated at CBS-QB3 level.[14,15])

(*Z*)-4-methylpent-2-ene. The most favorable conformation of the former (structure **B1**) points the two methyl substituents as far away from the C–C double bond as is possible in this small system (Figure 1.42b). The two substituents facing each other at the C1 and C3 positions of the allyl system (that is, the substituents pointing downwards in Figure 1.42b) are two C–H bonds. The second best conformation, **B2**, is energetically less favorable by 1.6 kJ mol^{-1} and positions one of the terminal methyl substituents pointing downwards. A third conformation, **B3**, exists at +8.1 kJ mol^{-1}, but can be characterized as a rotational transition state around the C2–C3 bond.

Replacing the terminal hydrogen in alkene **B** by a methyl group leads to (*Z*)-4-methylpent-2-ene (**C**) and a dramatic change in conformational energies. The best conformation **C1** is still the one pointing the two equivalent methyl substituents as far away as possible from the C–C double bond. Structure **C2**, where one of these methyl substituents points downwards, is energetically quite unstable at +16.2 kJ mol^{-1} and corresponds to a rotational transition state around the C3–C4 single bond. Relaxation of this structure leads to conformation **C3**, a true minimum on the potential energy surface located 13.8 kJ mol^{-1} above **C1**. Comparing the conformational energies of the second best conformers **B2** *vs.* **C2** we may conclude that A^{1,3} strain between two methyl substituents amounts to approximately 15 kJ mol^{-1} and thus represents a strong ordering force in otherwise flexible systems. In a more general sense we can conclude that in alkenes with (Z)-configuration the substituent present at one of the alkene carbon atoms controls the conformation of the substituent attached to the other alkene carbon atom.

1.6.2 Ring Strain

The strain energy contained in small- and medium-sized rings is essential in order to understand ring-forming or ring-breaking chemical transformations. One option for quantifying ring strain energies involves the prediction of a strain-free heat of formation for a given carbocycle using the Benson group increment for the C-(C)$_2$(H)$_2$ unit of -20.9 kJ mol^{-1} from Table 1.1. Comparison with the experimentally measured value then provides a quantitative way to assess ring strain. For the cycloalkanes ranging from cyclopropane to cyclooctane the respective results have been collected in Table 1.2. The most highly strained rings are cyclopropane and cyclobutane, with strain energies of $+116.0$ and $+111.9$ kJ mol^{-1}, respectively. The strain is much

Table 1.2 Heats of formation and ring strain energies derived from Benson group increments for selected cycloalkanes (data from ref. 12).

System	Structure	ΔH_f° (exp) (kJ mol^{-1})	ΔH_f°(GAV) (kJ mol^{-1})	Ring strain (kJ mol^{-1})
Cyclopropane	△	+53.3	−62.7	+116.0
Cyclobutane	□	+28.3	−83.6	+111.9
Cyclopentane	⬠	−76.4	−104.5	+28.1
Cyclohexane	⬡	−123.4	−125.4	+2.0
Cycloheptane	⬡	−118.1	−146.3	+28.2
Cyclooctane	⯃	−124.4	−167.2	+42.8

reduced in cyclopentane and reaches a value close to zero for cyclohexane ($+2.0$ kJ mol^{-1}). Increasing the ring size further then leads to moderately sized strain energies, as in cycloheptane ($+28.2$ kJ mol^{-1}) or cyclooctane ($+42.8$ kJ mol^{-1}).

An alternative approach for calculating ring strain energies in cycloalkanes uses acyclic alkanes as unstrained reference systems. In practice, this is done using **isodesmic reactions,** which have originally been defined as transformations conserving the number of bonds of a given formal type, but involve a change in how these bonds are interconnected (see ref. 16). An isodesmic reaction chosen to quantify the ring strain in cyclopropane is shown in eqn (1.5) and involves the formal reaction of cyclopropane with three equivalents of methane to yield three equivalents of ethane. On both the left and right sides of this reaction we count 18 C–H and 3 C–C bonds, thus satisfying the formal requirement of an isodesmic reaction. The reaction energy for this transformation, where the strained cyclopropane ring is transformed into unstrained product molecules (ethane), is $\Delta H_{298}(1.5) = -80.0$ kJ mol^{-1}, a value that is significantly smaller than that obtained using the Benson GAV approach.

$$\triangle \; + \; 3\ CH_4 \quad \xrightarrow{\Delta H_{298}} \quad 3\ H_3C{-}CH_3 \tag{1.5}$$

$$\triangle \; + \; 3\ H_3C{-}CH_3 \quad \xrightarrow{\Delta H_{298}} \quad 3 \; \wedge \tag{1.6}$$

A common criticism of isodesmic reactions is that reactants and products often differ in the types of bond, even if the formal requirement of "a given formal type" is satisfied. In eqn (1.5), for example, we are converting the 6 secondary C–H bonds in cyclopropane and the 12 C–H bonds in methane to 18 primary C–H bonds in the ethane products. A subclass of isodesmic reactions, termed **homodesmotic reactions,** attempts to remove this deficiency, and is defined as reactions involving, in reactants and products, equal numbers of carbon–carbon bonds connecting centers of a given formal hybridization (*e.g.* $C(sp^3){-}C(sp^3)$) and also equal numbers of carbon atoms of a given hybridization (*e.g.* $C(sp_3)$) with zero, one, two or three hydrogens attached.[17] A homodesmotic reaction transforming the strained cyclopropane molecule into unstrained products is shown in eqn (1.6), where we count (a) 6 $C(sp^3){-}C(sp^3)$ bonds and (b) 3 $C(sp^3)H_2$ and 6 $C(sp^3)H_3$ centers on either side of the equation. The reaction energy now is calculated as $\Delta H_{298}(1.6) = -116.0$ kJ mol^{-1}, which is in quantitative agreement with the value derived using the Benson group increment approach.

None of the approaches presented here for the quantification of ring strain reveal its origin. Analysis of the three-dimensional structures of the four smallest cycloalkanes, shown in Figure 1.43, is helpful in identifying various factors. The earliest rationalization of ring strain was based on bond angle deformations, often referred to as **Baeyer strain**. In cyclopropane, all three carbon atoms are located in one plane and the bond angles thus equate to 60°. This is obviously quite different from the bond angles of around 109° found in many unstrained alkanes and also in cyclohexane. Further inspection of the cyclopropane structure reveals another source of ring strain, namely the eclipsing orientation of all C–H bonds on adjacent carbon atoms. This is often referred to as **Pitzer strain**. Both types of strain are expected to appear in planar cyclobutane, where C–C–C bond angles of 90° and eclipsing C–H bonds are again expected to lead to a highly strained system. However, cyclobutane adopts a slightly puckered structure in order to reduce Pitzer strain to some extent. Planar cyclopentane with all carbon atoms in one plane allows favorable C–C–C bond angles of 108°, but would be plagued by a large number of eclipsing C–H interactions. The preferred conformation of cyclopentane is therefore that of an envelope with one of the five carbon atoms reaching out of the plane defined by the other four, which reduces the number of eclipsing C–H interactions and leads to slightly smaller bond angles. Cyclohexane prefers the well-known chair conformation, which allows for bond angles close to 109° and fully staggered C–H bonds. One final point concerns the almost identical ring strain energies in cyclopropane (+116.0 kJ mol^{-1}) and cyclobutane (+111.9 kJ mol^{-1}), a fact not easily reconciled with the apparently much larger bond angle distortions in the former compared with the latter. A more detailed analysis of cyclopropane reveals that the C–C bond orbitals do not follow the atom-to-atom connecting lines, but protrude outwards and so away from the ring center, as shown schematically in Figure 1.44a. This "bending" phenomenon does not, however, go as far as localizing

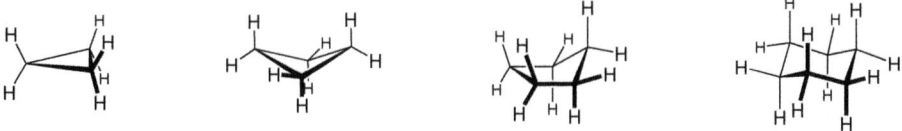

Figure 1.43 Structural representations for the four smallest cycloalkanes.

(a)   (b)

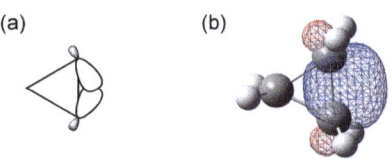

Figure 1.44   Schematic drawing of (a) bent C–C or "banana" bonds and (b) C–C bond NBOs in cyclopropane.

most of the C–C bond electron density outside the three membered ring, as is apparent from the C–C bond orbital calculated with the aid of the natural bond orbital (NBO) localization procedure (Figure 1.44b). In any case the bond angles in cyclopropane are systematically larger than 60°, which, in valence bond hybrid orbital terms, implies larger p-character in the orbitals used for C–C bond formation. At the same time this leaves more s-character for the description of C–H bonds and, consequently, higher C–H bond energies. These two (stabilizing) effects, together with additional factors present in cyclobutane, lead to almost identical strain energies in cyclopropane and cyclobutane.

1.7   Aromaticity

The molecular orbitals describing the π-system of aromatic systems, such as benzene following the Hückel $(4n + 2)$ rule, and those of anti-aromatic systems, such as cyclobuta-1,3-diene with $(4n)$ π-electrons, have already been discussed in detail in Section 1.3.2. While this type of analysis provides us with simple and easily applicable rules for differentiating aromatic and anti-aromatic π-systems, the actual degree of aromatic stabilization is not easily derived. In the following we will show how the aromatic stabilization energy of benzene, as the prototypical aromatic molecule, can be quantified using isodesmic and homodesmotic reactions. The earliest attempt to quantify aromatic stability was based on the comparison of hydrogenation enthalpies of aromatic and non-aromatic π-systems. The following discussion is based on thermochemical data reported in ref. 12. The hydrogenation of cyclohexene, for example, as expressed by eqn (1.7), is exothermic by $\Delta H_{298}(1.7) = -118.4$ kJ mol^{-1} and thus provides a reference for an isolated double bond located in a six-membered ring system. Hydrogenation of the three formal double bonds in benzene, as expressed by eqn (1.8) is exothermic by $\Delta H_{298}(1.8) = -206.0$ kJ mol^{-1}, a value significantly smaller than expected from the result for reaction (1.7).

The difference between what is expected for three isolated double bonds $(3 \times -118.4 = -355.2$ kJ mol$^{-1})$ and what is actually found for benzene amounts to $\Delta H_{298} = -355.2 + 206.0 = -149.2$ kJ mol^{-1} and is sometimes cited as the quantitative reflection of the stability of the aromatic π-system in benzene.

$$\text{\Large\bigcirc\mkern-20mu|} \quad + \quad H_2 \quad \xrightarrow{\Delta H_{298}} \quad \bigcirc \tag{1.7}$$

$$\bigcirc\mkern-14mu| \mkern-4mu| \quad + \quad 3 \ H_2 \quad \xrightarrow{\Delta H_{298}} \quad \bigcirc \tag{1.8}$$

$$3 \ \bigcirc\mkern-20mu| \quad \xrightarrow{\Delta H_{298}} \quad \bigcirc\mkern-14mu| \mkern-4mu| \quad + \quad 2 \ \bigcirc \tag{1.9}$$

The logic of calculating the **aromatic stabilization energy** (sometimes referred to as ASE) as outlined above can also be formulated as the transfer-hydrogen reaction shown in eqn (1.9). Comparing the left- and right-hand sides of this equation reveals that this reaction belongs to the isodesmic type, which proceeds with a reaction enthalpy of exactly $\Delta H_{298}(1.9) = -149.2$ kJ mol^{-1}. Formally, eqn (1.9) is obtained by taking three times eqn (1.7) and subtracting eqn (1.8). The ASE value obtained from eqn (1.9) includes other factors, in addition to the actual aromatic stabilization energy, such as the resonance energy obtained whenever isolated double bonds are coupled into a single conjugated π-system. Differentiation of these two effects is possible through the selection of reference systems containing conjugated C–C double bonds. One suitable choice appears to be cyclohexa-1,3-diene, whose hydrogenation as given in eqn (1.10) is exothermic by $\Delta H_{298}(1.10) = -229.6$ kJ mol^{-1}. Combination with eqn (1.8) describing the hydrogenation of benzene leads to the isodesmic eqn (1.11), which effectively describes the transfer-hydrogenation between three cyclohexa-1,3-diene molecules such that two benzene rings and one equivalent of cyclohexane are generated. This reaction is exothermic by $\Delta H_{298}(1.11) = -276.8$ kJ mol^{-1}, which implies an ASE value of -138.4 kJ mol^{-1} (due to the formation of two benzene rings in eqn (1.11)). One remaining deficiency of eqn (1.11) derives from the C(sp^3)–C(sp^2) bonds present on the reactant side, but not the product side. This difference can be removed by turning eqn (1.11) into homodesmotic eqn (1.12), which now involves equal numbers of C(sp^3)–C(sp^3), C(sp^3)–C(sp^2) and C(sp^2)–C(sp^2) bonds on either side of the equation.

This reaction is exothermic by $\Delta H_{298}(1.12) = -127.6$ kJ mol^{-1}, which represents an ASE value closely similar to that obtained from alternative approaches for quantifying the aromatic stabilization of benzene.

$$\text{(1.10)}$$

$$\text{(1.11)}$$

$$\text{(1.12)}$$

Following essentially the same logic, the anti-aromatic character of cyclobutadiene (CB) discussed already in Section 1.3.2 can be analyzed through comparison of its hydrogenation energy with that of a suitably chosen reference system. Hydrogenation of the single double bond in cyclobutene to cyclobutane is exothermic by $\Delta H_{298}(1.13) = -129$ kJ mol^{-1}. Complete hydrogenation of cyclobutadiene to cyclobutane is exothermic by $\Delta H_{298}(1.14) = -400 \pm 16$ kJ mol^{-1}. Comparison of this latter value with twice the hydrogenation energy of cyclobutene as given in eqn (1.13) yields a difference of 142 ± 16 kJ mol^{-1}. This comparison between the hydrogenation reactions in cyclobutene and cyclobutadiene can also be cast into eqn (1.15), where two molecules of cyclobutene react to give cyclobutadiene and cyclobutane with a reaction enthalpy of $\Delta H_{298}(1.15) = +142 \pm 16$ kJ mol^{-1}.[18] This value is sometimes referred to as the aromatic destabilization energy of cyclobutadiene, but we have to keep in mind that multiple factors might be at work here. In contrast to the six- membered ring systems shown in eqn (1.7)–(1.12), strain energies can be quite sizable in four-membered ring systems. Any change in strain energy between the reactants/products in eqn (1.15) may thus also surface in the large positive value of this hydrogen transfer reaction.

$$\text{(1.13)}$$

$$\text{(1.14)}$$

$$\text{(1.15)}$$

Structural data for aromatic and anti-aromatic π-systems have also been analyzed in an effort to quantify the degree of aromaticity. In qualitative terms, bond equalization between formal double and single bonds equates to stabilization through resonance delocalization of π-bonds. Reference values for "typical" C–C single and double bonds are those shown in Figure 1.45 for ethylene and ethane. The equilibrium bond distances (r_e) shown here are those for the vibrationless systems located at the minimum of the potential energy surface, which makes the data independent of temperature and isotopic substitution. Comparing the structure of buta-1,3-diene to those of ethylene and ethane, we see that resonance interaction between two C–C double bonds leads to minor lengthening of the formal C–C double bond, and a more significant shortening of the (formal) C–C single bond. Resonance interactions in benzene lead to full equalization of all C–C bonds located at 139.1 pm. The experimental structure of cyclobutadiene is not known, but the current best calculations predict a rectangular structure with formal double bonds of 134.9 pm and formal single bonds of 156.6 pm. The fact that the single bonds are actually longer than C–C single bonds in ethane may, in part, reflect a significant amount of ring strain in the system, but may also result from an effort to improve the stability of the π-system.

Experimental studies towards the structural characterization of cyclobutadiene are hindered by its extremely rapid dimerization. Efforts to slow down this unwanted reactivity involve the synthesis of derivatives carrying sterically demanding substituents, such as tetra-*tert*-butylcyclobutadiene (Figure 1.46). Starting from 2,3,4,5-tetra-*tert*-butylcyclopenta-2,4-diene-1-one, low temperature photolysis at 254 nm leads, through "criss-cross" valence tautomerization and subsequent elimination of carbon monoxide, to tetra-*tert*-butyltetrahedrane

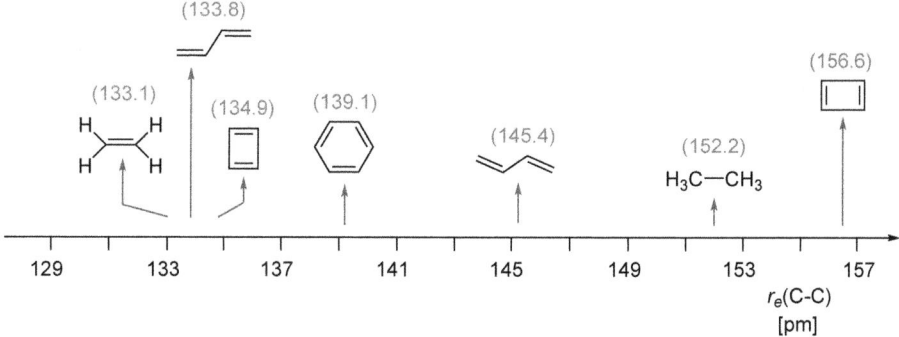

Figure 1.45 Equilibrium C–C bond distances (r_e, in pm) in selected systems.[19-21]

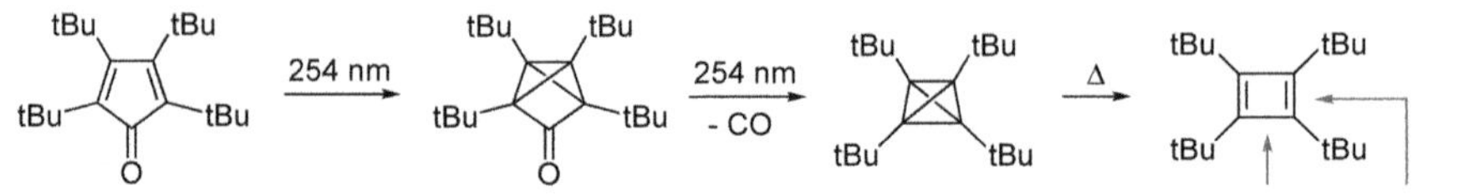

2,3,4-tetra-*tert*-butylcyclo-
penta-2,4-dien-1-one

T [°C]	$r(C\text{-}C)$ [pm]	$r(C{=}C)$ [pm]
20	148.2	146.4
-30	149.2	146.6
-150	152.6	144.2

Figure 1.46   Synthesis and structure of tetra-*tert*-butylcyclobutadiene.[22-24]

(among others). Thermal isomerization of this latter compound gives tetra-*tert*-butylcyclobutadiene as a thermally stable but air-sensitive product. X-ray crystal structure determination of its structure has been challenged by an unexpectedly strong temperature effect: while the "short" and "long" sides of the expected rectangular structure have almost identical lengths at room temperature and at −30 °C, with larger differences observed at −150 °C. The bond distance of 144.2 pm obtained for the short side of the rectangular structure is, however, still much longer compared to other C–C double bonds (see Figure 1.45 for comparison), and it thus remains unclear whether the dynamic processes responsible for the observed temperature effects are completely "frozen out", even at −150 °C.

References

1. *CRC Handbook of Chemistry and Physics*, ed. D. R. Lide, CRC Taylor & Francis, 89th edn, 2008.
2. D. H. Aue, in *Dicoordinated Carbocations,* ed. Z. Rappoport and P. J. Stang, Wiley, New York 1997, ch. 3, pp. 105–156.
3. H. Mayr and A. R. Ofial, in *Carbocation Chemistry*, ed. G. A. Olah and G. K. S. Prakash, Wiley, Hoboken, NJ, 2004, ch. 13, pp. 331–358.
4. Y.-R. Luo, *Comprehensive Handbook of Chemical Bond Energies*, CRC Press, 2007.
5. S. J. Angyal, *Angew. Chem., Int. Ed.*, 1969, **8**, 157.
6. F. Franks, P. J. Lillford and G. Robinson, *J. Chem. Soc., Faraday Trans. 1*, 1989, **85**, 2417.
7. E. L. Eliel and S. H. Wilen, *Stereochemistry of Organic Compounds*, John Wiley & Sons, 1994.
8. M. W. Chase Jr., *et al.*, NIST-JANAF Thermochemical Tables, 4th edn, *J. Phys. Chem. Ref. Data*, Monograph 9, 1998, pp. 1–1951.
9. L. Goodman, H. Gu and V. Pophristic, *J. Phys. Chem. A*, 2005, **109**, 1223.
10. L. Song, M. Liu, W. Wu, Q. Zhang and Y. Mo, *J. Chem. Theory Comput.*, 2005, **1**, 394.
11. Y. Mo, *Wiley Interdiscip. Rev.: Comput. Mol. Sci.*, 2011, **1**, 164.
12. N. Cohen, *J. Phys. Chem. Ref. Data*, 1996, **25**, 1411.
13. J. Csontos, B. Nagy, L. Gyevi-Nagy, M. Kallay and G. Tasi, *J. Chem. Theory Comput.*, 2016, **12**, 2679.

14. F. Johnson and S. K. Malhotra, *J. Am. Chem. Soc.*, 1965, **87**, 5492.
15. R. W. Hoffmann, *Chem. Rev.*, 1989, **89**, 1841.
16. W. J. Hehre, R. Ditchfield, L. Radom and J. A. Pople, *J. Am. Chem. Soc.*, 1970, **92**, 4796.
17. S. E. Wheeler, K. N. Houk, P. v. R. Schleyer and W. D. Allen, *J. Am. Chem. Soc.*, 2009, **131**, 2547.
18. A. Fattahi, L. Lis, Z. Tian and S. R. Kass, *Angew. Chem., Int. Ed.*, 2006, **45**, 4984.
19. M. Piccardo, E. Penocchio, C. Puzzarini, M. Biczysko and V. Barone, *J. Phys. Chem. A*, 2015, **119**, 2058.
20. N. C. Craig, P. Groner and D. C. McKean, *J. Phys. Chem. A*, 2006, **110**, 7461.
21. S. V. Levchenko and A. I. Krylov, *J. Chem. Phys.*, 2004, **120**, 175.
22. G. Maier, S. Pfriem, U. Schäfer and R. Matusch, *Angew. Chem., Int. Ed.*, 1978, **17**, 520.
23. H. Irngartinger, N. Riegler, K.-D. Maksch, K.-A. Schneider and G. Maier, *Angew. Chem., Int. Ed.*, 1980, **19**, 211.
24. H. Irngartinger and M. Nixdorf, *Angew. Chem., Int. Ed.*, 1983, **22**, 403.

2 Reactivity Models in Organic Chemistry

2.1 Potential Energy Surfaces and How They Connect to Chemical Reactivity

Organic reaction mechanisms are commonly discussed with reference to qualitatively drawn **potential energy surfaces** (PESs), their qualitative nature simply being due to the fact that most of the data required for a quantitative description are not available. Taking the reaction of chlorine radical and methane to give HCl and methyl radical shown in Figure 2.1 as an example, the reaction is initiated through formation of a weakly bound **reactant complex** (RC). From here the reaction proceeds through a **transition state** (TS), characterized by simultaneous C–H bond breaking and Cl–H bond making, down to the **product complex** (PC), which ultimately dissociates to the HCl and methyl radical products. Reaction progress is described by a reaction coordinate, whose actual definition is normally not given and whose relationship to structural parameters, such as C–H or Cl–H bond distances, varies along the reaction pathway. The minima (RC, PC) and maxima (TS) shown in Figure 2.1 differ from all other points in the reaction profile in that the first derivative of the potential energy with respect to the reaction coordinate is zero at these positions. These so-called **stationary points** are thus located either on a valley floor or on a hilltop. However, this is true only with respect to the reaction coordinate, and all stationary points are in a minimum location with respect to all other structural variables. Any non-linear molecular system composed of N

Reactivity and Mechanism in Organic Chemistry, 2nd Edition
By Hendrik Zipse
© Hendrik Zipse 2023
Published by the Royal Society of Chemistry, www.rsc.org

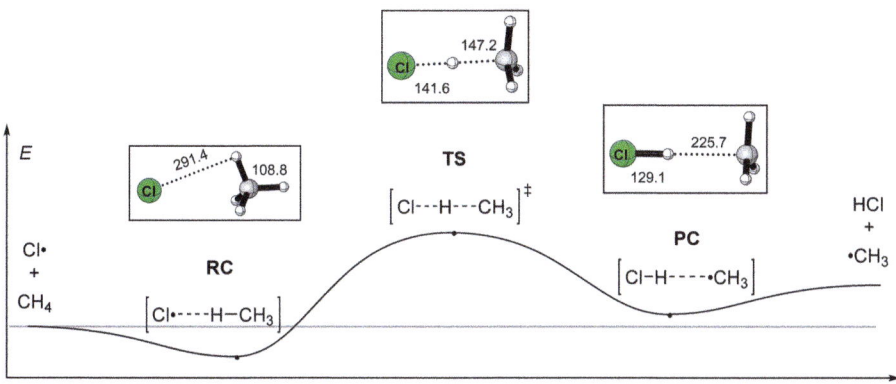

Figure 2.1 Qualitative potential energy surface for the reaction of chlorine radical with methane. Distances are given in pm.

atoms, such as the one shown in Figure 2.1, has $3N - 6$ structural degrees of freedom in internal coordinate space, and for the system chosen here this yields $3 \times 6 - 6 = 12$ internal degrees of freedom. Aside from the reaction coordinate, this leaves us with 11 degrees of freedom (C–H bond distances, H–C–H bond angles, *etc.*) with respect to which the system resides in an energy minimum at all points along the reaction pathway. This is more readily appreciated when the PES is drawn as a two-dimensional (2D) energy diagram defined by the r(Cl–H) and r(H–C) bond distances, adding the energy information through contour lines, as shown in Figure 2.2 for the same reaction as in Figure 2.1. The reactant complex RC is located in the upper left corner of the diagram. From here the reaction pathway of lowest energy (the **minimum energy reaction pathway**, MERP or MEP) runs in the direction of shorter Cl–H distances towards the transition-state region. Just before reaching the transition state the MEP takes a turn and is then defined by simultaneously shortening Cl–H and lengthening H–C distances. After passing through the TS, the MEP proceeds downwards towards the product complex PC, a process mainly described by increasing H–C bond distances. When choosing the MEP as the reaction pathway, the reaction is moving along the valley floor even when passing through the transition state. This can be visualized by potential energy surface cuts orthogonal to the MEP, as shown in the three insets to the right in Figure 2.2. The first cut (1) shows how reactant complex RC is located at the bottom of the PES valley, both along the MEP as well as in the orthogonal direction. The second cut (2) shows how the TS represents an energy minimum in a direction orthogonal to the MEP, the only notable difference to

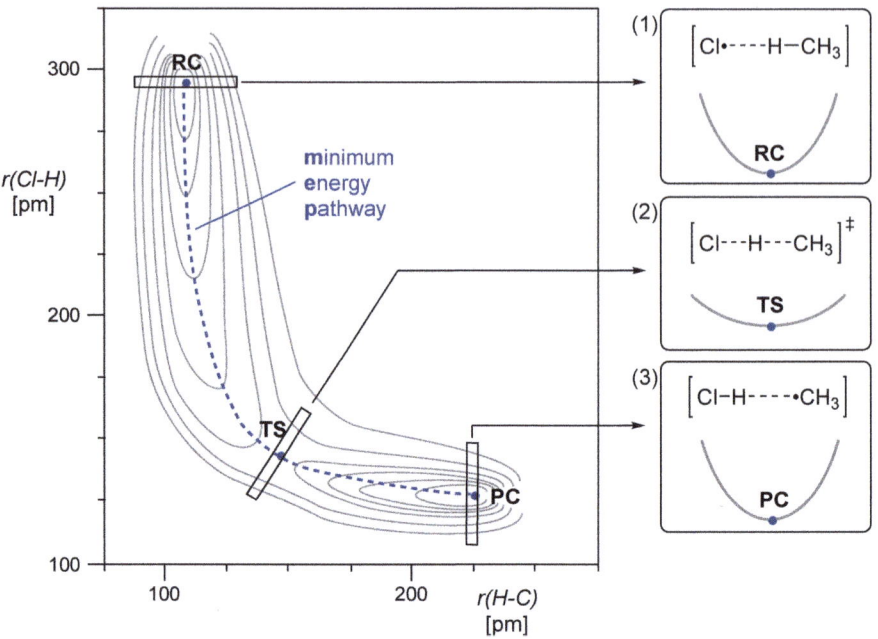

Figure 2.2 2D energy diagram for the reaction of chlorine radical with methane.

the region around the reactant complex being the flatter shape of the PES. This implies that the TS is a saddle point, in that it represents an energy maximum along the MEP, but an energy minimum along all other geometrical degrees of freedom. The last cut (3) for the PC shows the same situation as for the RC, in that this stationary point is a minimum with respect to both degrees of freedom chosen here.

The outcome of molecular reactions is not only governed by the characteristics of local minima and transition states, but also, in some cases, is influenced by what is commonly referred to as a **valley ridge inflection** (VRI) point. The VRI marks the position where a product valley coming down from the transition-state region divides (or *bifurcates*) into two separate product valleys. One of the best-studied cases for this situation concerns the isomerization of the methoxy radical (**A**) to the much more stable hydroxymethyl radical (**B**) (Figure 2.3). In principle this is a fairly simple process, where one of the three methyl group hydrogen atoms migrates from the carbon to the oxygen atom.

The C_s-symmetric transition state for this process is characterized by a symmetry plane defined by the migrating hydrogen, the carbon and the oxygen atoms. Moving down from transition state **TS1** towards the product region in C_s symmetry, the system eventually arrives at **TS2**,

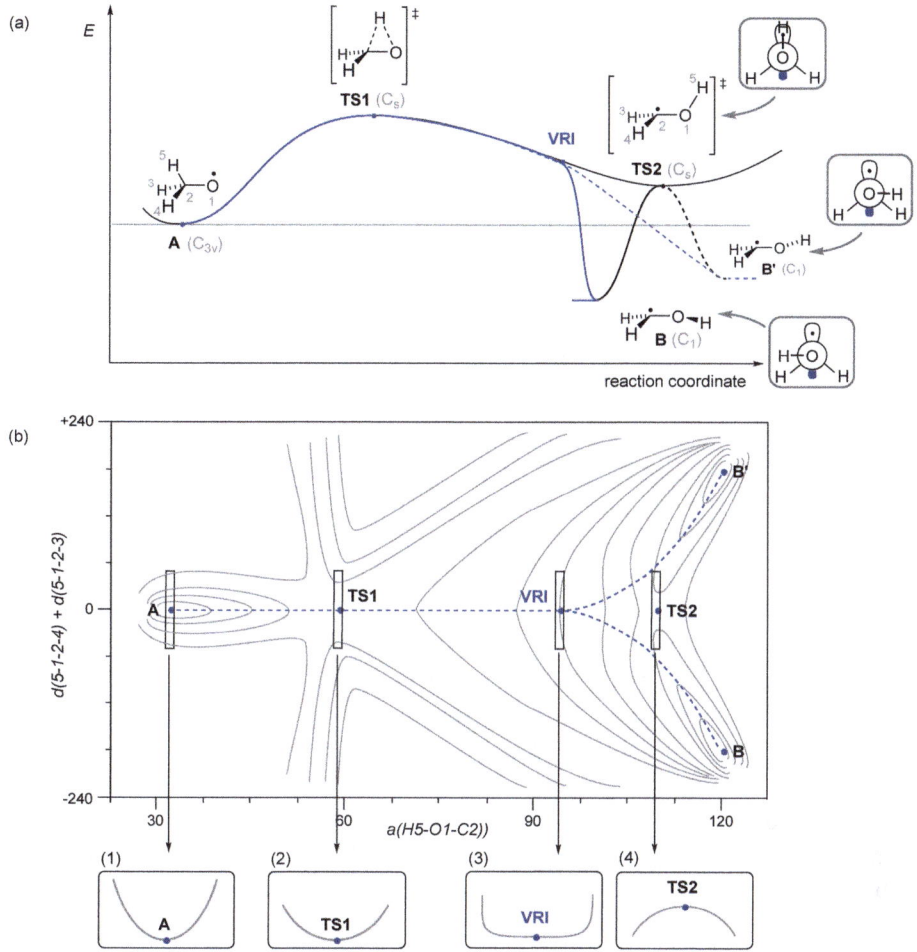

Figure 2.3 (a) Schematic reaction pathway and (b) 2D energy diagram for the isomerization of methoxy radical (**A**, C_{3v}) to hydroxymethyl radical (**B**, C_1).

which describes the symmetry-breaking rotation around the central C–O bond. Moving down from **TS2** to either side of the PES leads to C_1-symmetric hydroxymethyl radicals with either **B** or **B'** structure as the final reaction products. As is readily seen from the Newman projections in Figure 2.3a, these two species differ only in whether the O–H group hydrogen moves clockwise (**B'**) or counterclockwise (**B**) when descending from **TS2**. Interestingly, the most favorable reaction pathway does not lead all the way to **TS2**, but diverges from the symmetric structure at the VRI. The nature of this point can readily be appreciated by again following the cuts taken through the PES orthogonal to the MEP: while the hydrogen migration **TS1** is located in an energy minimum, the VRI represents a point of inversion in this direction, while the C–O bond rotation **TS2** represents an energy maximum. Although the isomerization reaction chosen here as an example may not seem overly relevant for the area of organic synthesis, bifurcations of reaction pathways such as the one found here occur quite frequently in reactions where highly symmetric reactants transform to products of lower symmetry.

A final comment on the reaction between chlorine radical and methane shown in Figure 2.1 must be made concerning the actual definition of the energy "*E*". This is the potential energy of the systems at 0 K in an artificial (classical) vibrationless state. The advent of powerful computational chemistry algorithms implemented in efficient computational chemistry software packages has made it possible to compute potential energy surfaces for a wide variety of organic reactions and thus test most of the reaction mechanisms presented in introductory organic chemistry textbooks. Still, the potential energy that is defined classically is far from energies that can be tested experimentally. First of all, this is due to the fact that molecular systems are never at rest, but have **zero point vibrational energy** (ZPVE or ZPE), even at 0 K. In addition, most experiments on molecular systems are not performed at cryogenic temperatures, but usually refer to the most often employed "standard state" conditions (temperature of 298.15 K, 1 bar pressure). Correcting for all these differences leads to the enthalpy at 298.15 K (H_{298}) gas-phase energy definition for molecular systems that is most often employed. For the reactants and products shown in the reaction in Figure 2.4, accurate experimental data are available and we can therefore verify that the overall reaction enthalpy calculated with quantum mechanical methods (ΔH_{298} = +9.8 kJ mol^{-1}) is quite close to that calculated from experimental enthalpy data (ΔH_{298} = +7.4 kJ mol^{-1}) in this case. With respect to the reaction profiles, the differences between E and H_{298} are most pronounced when comparing transition states with their adjacent minima, as the zero-point energies of the former are typically lower than those of the latter. This is readily seen in Figure 2.4, where a potential energy barrier of ΔE = +29.1 kJ mol^{-1} is reduced to an enthalpy barrier of ΔH_{298} = +7.1 kJ mol^{-1}. Reaction rates are governed by Gibbs free energies G_{298} that differ from enthalpies by the additional consideration of **entropic effects,** according to $\Delta G = \Delta H - T\Delta S$. This equation is sometimes referred to as the **"Gibbs–Helmholtz equation"**. The differences between ΔH_{298} and ΔG_{298} are particularly large where the number of molecular species changes, which can readily be seen in Figure 2.4

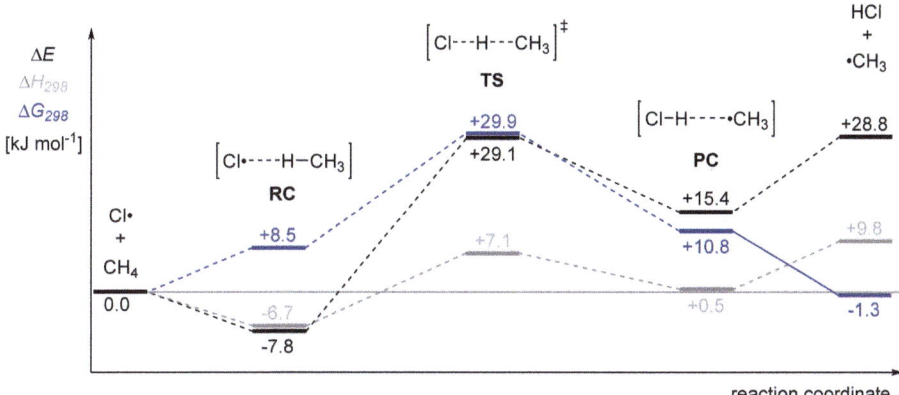

Figure 2.4 Reaction profiles for the reaction of chlorine radical with methane in terms of the potential energy (*E*), the enthalpy at 298.15 K (H_{298}) and the Gibbs free energy at 298.15 K (G_{298}) ((U)M06-2X/cc-pVTZ results, in kJ mol^{-1}).

when comparing energies for the formation of reactant and product complexes from their individual components: while reactant complex RC is more favorable than chlorine radical and methane in ΔH_{298} terms, the opposite is true with respect to ΔG_{298}. Reactant and product complexes, at least when only loosely bound as in the example here, thus play no significant role in free energy profiles of most reactions in organic chemistry, and so are often not even mentioned as relevant stationary points in mechanistic discussions.

With the free energy profile of a reaction to hand, we can employ **transition state theory** (developed by H. Eyring) to predict the corresponding reaction rate. One of the key assumptions made in this theory is that the transition state exists in a quasi-equilibrium characterized by equilibrium constant $K^{\ddagger}$ with the contributing reactants, and that the forward reaction rate starting from the transition state can be described with the universal rate constant $k^{\ddagger}$. For the sample reaction from Figure 2.4 this leads to the simplified description of the reaction mechanism as defined in eqn (2.1) and the assumption that the rate of reaction can be described by the rate eqn (2.2). Together with additional assumptions, the constants $K^{\ddagger}$ and $k^{\ddagger}$ can be developed into the **Eyring equation** (eqn (2.3)) for the calculation of the respective bimolecular rate constant k. Key ingredients of this expression are the activation free energy $\Delta G^{\ddagger}$, taken as the free energy difference between separate reactants and the transition state, and the reaction temperature T (in K). The transmission coefficient simply reflects the proportion of the (quasi-equilibrated) transition state moving forward towards the products (and not back to the reactants) and is often assumed to be unity.

$$Cl\cdot + CH_4 \overset{K^{\ddagger}}{\rightleftharpoons} \left[Cl\cdots\overset{TS}{H}\cdots CH_3 \right]^{\ddagger} \overset{k^{\ddagger}}{\longrightarrow} HCl + \cdot CH_3 \qquad (2.1)$$

$$r = k[Cl\cdot][CH_4] \qquad (2.2)$$

$$k = \kappa \, \frac{k_B T}{h} \exp(\frac{-\Delta G^{\ddagger}}{RT}) \qquad (2.3)$$

rate constant · Boltzmann constant · absolute temperature · activation free energy · transmission coefficient · Planck constant · gas constant

$$\ln k = \ln\left(\frac{k_B}{h}\right) + \ln T - \frac{\Delta G^{\ddagger}}{RT} \qquad (2.4)$$

$$\lg k = \lg\left(\frac{k_B}{h}\right) + \lg T - \frac{\Delta G^{\ddagger}}{2.303RT} \qquad (2.5)$$

$$\ln k = \ln\left(\frac{k_B}{h}\right) + \ln T - \frac{\Delta H^{\ddagger}}{RT} + \frac{\Delta S^{\ddagger}}{R} \qquad (2.6)$$

activation enthalpy · activation entropy

With this latter assumption and also taking the logarithmic form, we arrive at eqn (2.4). In some cases using the decadic logarithm form, as in eqn (2.5) provides some numerical advantages. Separating the free energy into its enthalpy and entropy components, as in eqn (2.6), then yields the most frequently used version of the Eyring equation, which is best suited to analyze the temperature dependence of reaction rates on a quantitative basis. This is also the goal of the **Arrhenius equation** (eqn (2.7)), where the pre-exponential factor A and the (Arrhenius) activation energy E_a are used for the calculation of rate constants, but this equation lacks the conceptual basis of transition state theory.

$$\ln k = \ln A - \frac{E_a}{RT} \qquad (2.7)$$

pre-exponential factor · activation energy

One of the interesting characteristics of eqn (2.3) is that it predicts an upper limit for rate constants for the simple case that the activation free energy equates to zero. For the hypothetical case of a **unimolecular reaction,** and assuming a reaction temperature of 298.15 K, this leads to $k_{max} = 6.2 \times 10^{12}$ s^{-1}, which is similar to the frequencies of stretching vibrations of bonds in organic molecules. The prediction is more difficult for the case of a **bimolecular reaction,** such as the one shown in eqn (2.1) because the rate of very fast reactions may also depend on how fast the two reactants meet. This is not accounted for in the simplified mechanistic expression of eqn (2.1) used for the derivation of the Eyring equation, where the rate constants for reactant association are hidden behind the quasi-equilibrium constant $K^{\ddagger}$. A more appropriate mechanistic ansatz is given in eqn (2.8), where the formation of the reactant complex through dissociative encounter of the reactant is now described by rate constant k_D and its dissociation back to reactants by k_{-D}. Using this expression, the forward reaction from the reactant complex to the products now corresponds to a unimolecular process described by rate constant k_1. Adding the assumption that the concentration of the reactant complex is not changing too much during the overall reaction (that is, it is quasi-stationary), the rate equation (eqn (2.9)) can be derived, where the observed overall rate constant (k_{obs}) is a function of the microscopic rate constants k_1, k_D and k_{-D}. For the case that the unimolecular rate constant k_1 is much larger than k_D and k_{-D}, this equation reduces to eqn (2.10), which implies that diffusive encounter of the two reactants becomes rate limiting once the barriers for the subsequent substrate conversion become negligible. Diffusion rate constants in organic solvents at 298 K are of the order of $k_D = 10^9 - 10^{10}$ l mol^{-1} s^{-1}, and **diffusion-controlled reactions** cannot be faster than this limit.

$$Cl\bullet + CH_4 \underset{k_{-D}}{\overset{k_D}{\rightleftharpoons}} \left[Cl\bullet \cdots \overset{RC}{H} - CH_3 \right] \overset{k_1}{\longrightarrow} HCl + \bullet CH_3 \qquad (2.8)$$

$$r = k_{obs}\left[Cl\bullet\right]\left[CH_4\right] = \frac{k_1 k_D}{k_1 + k_{-D}}\left[Cl\bullet\right]\left[CH_4\right] \qquad (2.9)$$

$$r = k_D[Cl\bullet][CH_4] \quad (\text{for } k_1 \gg k_D, k_{-D}) \qquad (2.10)$$

True elementary reactions defined through a transition state connecting a single reactant (complex) with a single product (complex) are directly linked to a rate equation, where the **reaction order** (the sum of the exponents of the reactant concentrations) is identical to the **molecularity** of the reaction: **unimolecular** reactions lead to **first-order** rate equations, and **bimolecular** reactions lead to **second-order** rate equations. For most of the reactions relevant in organic synthesis, which are composed of a series of elementary reactions steps, the direct connection between molecularity and reaction order cannot be made. Among the more complex reaction sequences, the two-step reaction shown in Figure 2.5 is encountered quite frequently and involves **pre-equilibrium** activation of one of the reactants, followed by rate-limiting transformation of this reactant in a second step. The free energy profile for such a sequence is shown in Figure 2.5a and combines a first step, with a low barrier, with a second step with a much higher barrier. This is a very common scenario for many acid/base-catalyzed reactions, where the substrates are activated through initial protonation (or deprotonation) steps. The corresponding acids/bases of the initial substrates thus formed then undergo further reactions in the follow-up step. The same sequence also applies to Lewis acid/base-catalyzed reactions, where substrate activation through reaction with the catalyst involves covalent bond formation. An example of this type of Lewis base catalysis is shown in Figure 2.5b and involves pre-equilibrium formation of intermediate **C** through reaction of benzoyl chloride (**A**) with pyridine derivative **B**. This step converts the neutral electrophile **A** into the cationic electrophile **C**, which accelerates the reaction, with alcohol **D** acting as a nucleophile. The HCl by-product formed in the last step is neutralized by NEt$_3$ (**F**) added as an auxiliary base. This mechanism is supported by the fact that substrate/catalyst adduct **C** can be identified by ^{1}H NMR spectroscopy.

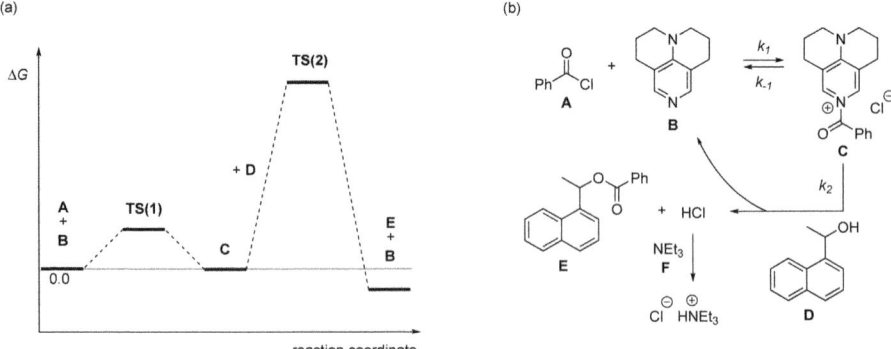

Figure 2.5   (a) Qualitative free energy profile and (b) reaction mechanism for the Lewis base-catalyzed reaction of benzoyl chloride (**A**) with 1-naphthylethanol (**D**).[1]

A second important two-step reaction combines a first step with a high barrier with a second, low barrier, step. This leads to the reaction profile shown in Figure 2.6a. The most prominent example for this type of reaction cited in many textbooks is the S_N1 substitution reaction of tertiary halides (**G**), where the first (unimolecular) step generates an anionic leaving group together with a tertiary carbocation intermediate (**H**). Reaction of this latter species with an added nucleophile (**I**) or the solvent is usually fast and generates adduct **J** as the final reaction product.

Everything we have said so far in this chapter makes the silent assumption that the outcome of a reaction is determined by the rate constants for whatever happens to be the slowest step of the overall reaction. This is only true as long as the reactions are **kinetically controlled**, but is not the case for reactions with comparable forward and backward reaction rates (that is, under equilibrating conditions), where the reaction outcome is dominated by thermodynamic factors. The reaction is then said to be under **thermodynamic control**. This is most often found in reactions lacking a strong thermodynamic driving force. In the example shown in Figure 2.7a deprotonation of methyl isobutyrate by lithium diisopropylamide (LDA) at −78 °C in tetrahydrofuran (THF) generates a lithium enolate intermediate, whose addition to cyclohex-2-ene-1-one can occur at the C1 (carbonyl) or the C3 (alkene terminus) carbon atom. When the reaction mixture is quenched with water carefully (that is, without warming) at −78 °C, the major product isolated is the 1,2-addition product. In contrast, when the reaction mixture is allowed to warm to +25 °C before quenching with water, the major product is the 1,4-addition product.

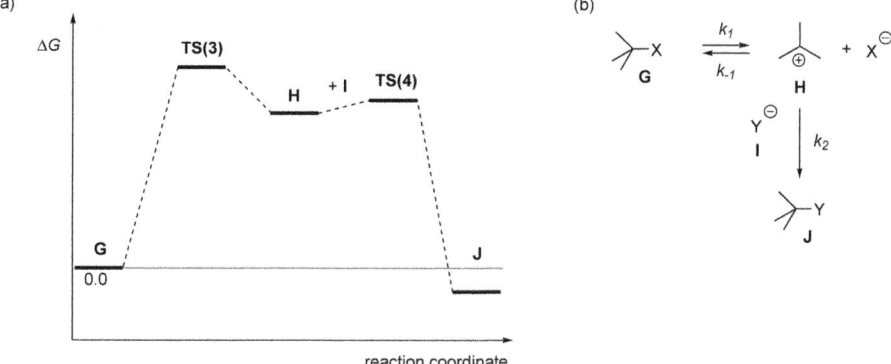

Figure 2.6 (a) Qualitative free energy profile and (b) reaction mechanism for the S_N1 substitution reaction of tertiary alkyl halides.

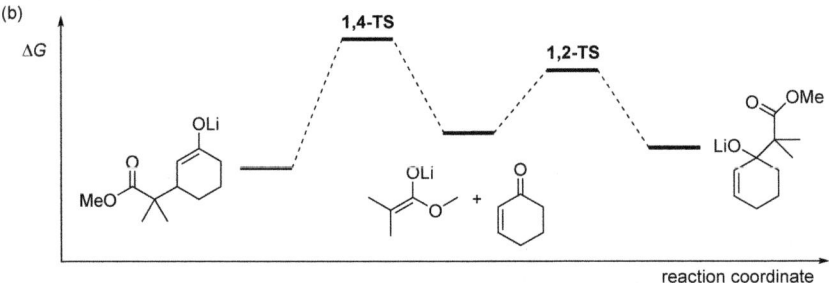

Figure 2.7 (a) Temperature effects in the 1,2- *vs.* 1,4-addition to cyclohex-2-ene-1-one and (b) qualitative free energy profile for the enolate addition steps.[2]

The explanation for the strong temperature dependence of the product mixture rests on the assumption that enolate addition to the carbonyl carbon atom is faster, but generates the less stable adduct compared with the addition to the C3 position. In the free energy profile shown in Figure 2.7b, this corresponds to a lower barrier for 1,2- than for 1,4-addition. Quenching at low temperature thus generates a product mixture reflecting the initial rate of enolate addition to cyclohex-2-ene-1-one (kinetic control). On warming to +25 °C, the initially formed enolate adducts start to equilibrate, most likely through reversion to the two reactants, followed by re-addition. Quenching at higher temperature thus gives a product mixture reflecting the relative thermodynamic stabilities of the two enolate adducts (thermodynamic control). As is the case for this particular example, the actual (relative) stabilities of regio- or stereoisomeric products are normally not known.

One of the few exceptions concerns the addition of thiocyanate to substituted benzhydryl cations (Figure 2.8). Due to its ambient nature, thiocyanate can add with its S- or N- terminus to the cation center, which yields thiocyanates or isothiocyanates as products. The experimentally measured reaction rates for S-addition are more than three orders of magnitude larger than for N-addition, a result also found for other highly stabilized cations. In contrast to the high kinetic preference for S-addition, the only products isolated from the reaction mixture are the isothiocyanates obtained through N-addition. Analysis of the actual equilibrium constants for the two addition channels, and thus the Gibbs free energies of addition, clearly show that the isothiocyanate products are more stable than the thiocyanates by approximately 10 kJ mol^{-1}. At the stage of product isolation, the reaction is therefore under thermodynamic control.

From a general point of view all organic reactivity phenomena discussed in this chapter can be cast into the "reactivity tree" shown in Figure 2.9, where the first branching point concerns the question of kinetic *vs.* thermodynamic control. For kinetically controlled reactions a further division can be made into reactions showing diffusion-limited reactivity and those that have a true enthalpic reaction barrier (activation control). Only for this last class of reactions can the

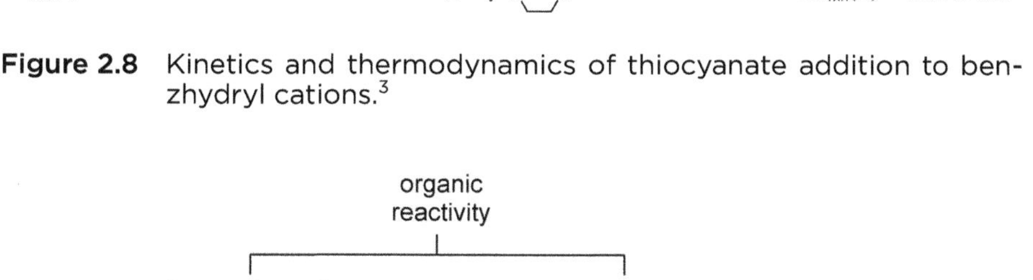

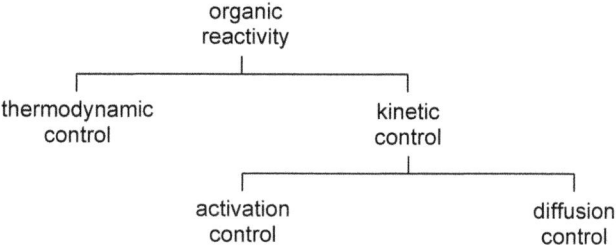

Figure 2.8 Kinetics and thermodynamics of thiocyanate addition to benzhydryl cations.[3]

Figure 2.9 General classification of organic reactivity phenomena.

reactivity models described in more detail in the following sections be applied, such as the Bell–Evans–Polanyi principle, Marcus theory, the hard and soft acids and bases (HSAB) principle, the Mayr–Patz equation or frontier molecular orbital theory.

2.2 The Bell–Evans–Polanyi Principle

One of the oldest reactivity models in organic chemistry goes back to conceptual work by R. P. Bell, M. G. Evans and M. Polanyi, and can most readily be explained with reference to the radical substitution reaction shown in eqn (2.11). The reaction enthalpy for this reaction is simply the bond energy difference between the B–C bond on the reactant side and the A–B bond on the product side. The origin of the reaction barrier $\Delta H^{\ddagger}(A)$ for this reaction can be described as the result of the intersection of the B–C and A–B potential energy curves as shown pictorially in Figure 2.10.

The horizontal placement of these curves relative to each other is arbitrary to some extent, but the vertical displacement is accurately given by the reaction enthalpy $\Delta H(A)$. Let us now compare this reaction with a closely related second case, where reaction of radical •A′ with B–C leads in a more exothermic way to the same product radical •C and a new substitution product A′–B. In Figure 2.10a this will involve the same potential energy curve for the reactant B–C, but a new curve for product A′–B. Due to the increased exothermicity of the reaction, this curve is shifted downwards relative to the curve for product A–B.

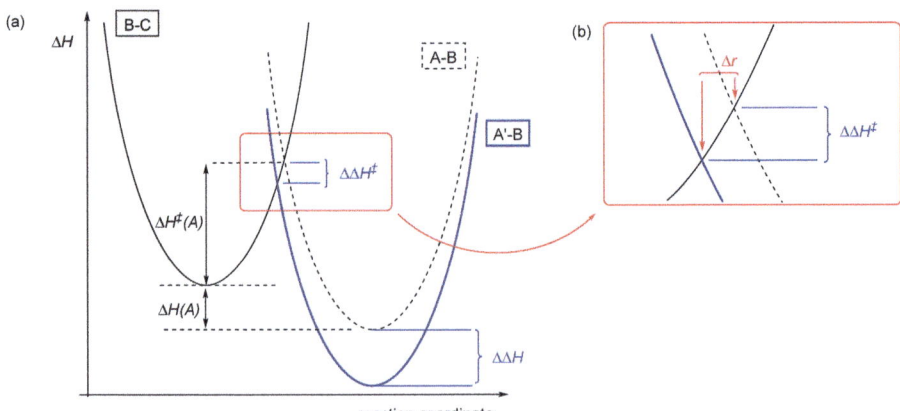

Figure 2.10 (a) Intersecting reactant and product potential energy curves illustrating the Bell–Evans–Polanyi principle. (b) Close-up view of the crossing region of the potential energy curves.

The interesting result of this shift is that the position of the intersection with the reactant curve also shifts to lower energies, predicting a lower activation energy for the reaction of radical •A′ relative to radical •A. So, in short, the **Bell–Evans–Polanyi (BEP) principle** states that more exothermic reactions lead to lower barriers. This statement is obviously only meaningful when comparing two mechanistically closely related reactions. A quantitative expression for the BEP principle is given by eqn (2.13), where differences in reaction barriers $(\Delta\Delta H^{\ddagger})$ relate to differences in the associated reaction enthalpies $\Delta\Delta H$ with a scaling factor α. The actual magnitude of this scaling factor is often found to vary around 0.5, but may also assume significantly smaller values.

$$\text{A} \bullet + \text{B} - \text{C} \xrightarrow{\Delta H^{\ddagger}(\text{A})} \text{A} - \text{B} + \text{C} \bullet \tag{2.11}$$

$$\text{A}' \bullet + \text{B} - \text{C} \xrightarrow{\Delta H^{\ddagger}(\text{A}')} \text{A}' - \text{B} + \text{C} \bullet \tag{2.12}$$

$$\Delta\Delta H^{\ddagger} = \alpha\Delta\Delta H \tag{2.13}$$

In the close-up view of the crossing region of the curves shown in Figure 2.10b, we can also see that shifting the product curves relative to each other *vertically* leads to a change in the location of the crossing point (Δr) *horizontally*. That the more exothermic reaction apparently has a geometrically earlier transition state is sometimes referred to as the **Hammond** or **Hammond–Leffler postulate**. However, direct access to structural data for transition states is extremely difficult and validation of this postulate therefore meets with substantial difficulties.

2.3 Marcus Theory

According to the Marcus model the barrier for a bimolecular reaction results from the intersection of parabolic reactant and product energy curves. For identity reactions, such as the reaction of chloride ion with methyl chloride, this gives rise to a reaction barrier called the **intrinsic reaction barrier** $\Delta G_0^{\ddagger}$ at the crossing point of the two potential curves (Figure 2.11a). For a non-identity reaction, such as the reaction of fluoride with methyl chloride, the intrinsic barrier is combined with two terms, as expressed in the **Marcus equation** (eqn (2.14)) describing the influence of the reaction energy ΔG° on the barrier $\Delta G^{\ddagger}$:

$$\Delta G^{\ddagger} = \Delta G_0^{\ddagger} + 0.5\Delta G^{\circ} + \frac{(\Delta G^{\circ})^2}{16\Delta G_0^{\ddagger}} \tag{2.14}$$

$$\Delta G_0^{\ddagger}(\text{AXB}) = 0.5[\Delta G_0^{\ddagger}(\text{AXA}) + \Delta G_0^{\ddagger}(\text{BXB})] \tag{2.15}$$

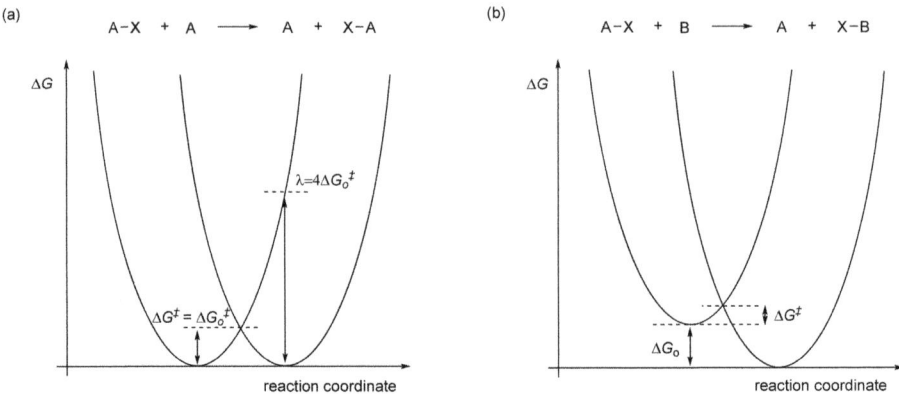

Figure 2.11 Intersecting reactant and product energy curves for (a) an identity reaction lacking a driving force and (b) a reaction with negative reaction free energy ΔG°.

The intrinsic barriers for non-identity reactions can often be approximated using the mixing rule (eqn (2.15)), where $\Delta G_o^\ddagger(\text{AXA})$ and $\Delta G_o^\ddagger(\text{BXB})$ refer to the intrinsic barriers for the individual identity reactions. The last term in eqn (2.14) is often quite small and the Marcus equation therefore predicts that approximately half of the reaction energy ΔG° enters into the reaction barrier $\Delta G^\ddagger$. This important relationship between reaction barrier (kinetics) and reaction energy (thermodynamics) implies that low reaction barriers $\Delta G^\ddagger$ can result from either low intrinsic reaction barriers $\Delta G_o^\ddagger$ or large negative reaction energies (or both).

The Marcus equation (eqn (2.14)) is often used to analyze the origin of reaction barriers. The reaction of enolate ion **A** with chloromethane **B** can, for example, lead to methyl vinyl ether **C** or propanal **D** (Figure 2.12).

The gas-phase barrier for *O*-alkylation amounts to $\Delta G^\ddagger(\text{O}) = +22.0$ kJ mol^{-1}, while that for *C*-alkylation is significantly higher at $\Delta G^\ddagger(\text{C}) = +41.1$ kJ mol^{-1}. The lower barrier for *O*- than for *C*-alkylation is in remarkable contrast to the much more favorable reaction energy for the latter ($\Delta G^\circ(\text{C}) = -160.0$ kJ mol^{-1}) than the former ($\Delta G^\circ(\text{O}) = -78.4$ kJ mol^{-1}). How, then, does the kinetic preference for *O*-alkylation arise in this case? Inserting barrier and reaction energy data into the Marcus equation yields intrinsic reaction barriers of $\Delta G_o^\ddagger(\text{O}) = +54.1$ kJ mol^{-1} and $\Delta G_o^\ddagger(\text{C}) = +106.6$ kJ mol^{-1}. Reaction at the O terminus of enolate **A** thus faces the much lower intrinsic barrier and the more favorable reaction energy for reaction at the C terminus cannot compensate for this intrinsic reactivity advantage. Intrinsic reaction barriers for ambident nucleophiles actually decrease systematically when moving from

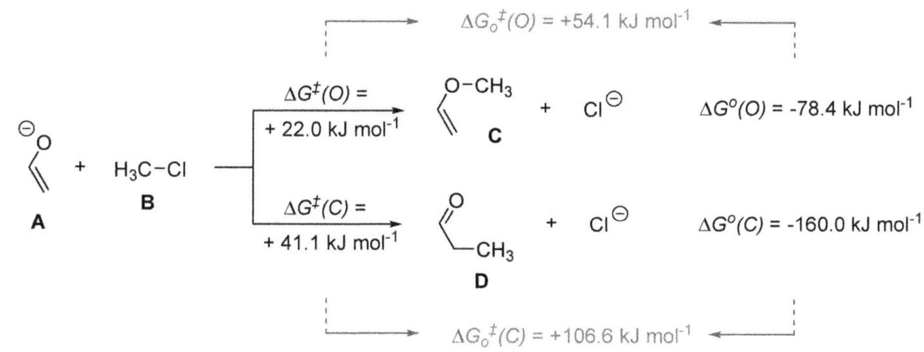

Figure 2.12 *O*- and *C*-Alkylation of enolate ion **A** with chloromethane **B**.[4]

carbon to the right and/or down in the periodic table. Reactions of enolate ions thus have systematically lower intrinsic barriers for reaction at the O terminus compared with the C terminus. The outcome of the reaction is then very much dependent on the reactant and its ability to generate sufficient reaction energy and thus compensate the intrinsic preference for reaction at the O terminus.

2.4 The Theory of HSAB

The theory of Hard and Soft Acids and Bases (**HSAB**) basically states that hard acids prefer to react with hard bases, as this maximizes ionic bond strength in the respective products, while soft acids prefer to react with soft bases, as this leads to products with maximum covalent bond strength. The acid/base terms used here follow the Lewis definition, and it is also clear from the above statement that the HSAB principle aims at the prediction of reaction thermodynamics. Although it may be tempting to apply HSAB arguments to kinetically controlled reactions also, there is actually no conceptual foundation for this practice. HSAB theory has its theoretical foundations in conceptual DFT. At least for discrete molecular and atomic species the energy of a system S can be defined as a function of the overall number of electrons N according to eqn (2.16).

$$E_N = E_{N0} - \chi_S(N - N_0) + \eta_S(N - N_0)^2 \tag{2.16}$$

This expression refers to a reference state of energy E_{N0} carrying N_0 electrons and uses the **Mulliken absolute electronegativity** (χ_S, greek "chi" symbol) and the **absolute hardness** (η_S, greek "eta" symbol) as parameters. These two quantities are accessible through the **ionization potential** (IP_S) and the **electron affinity** (EA_S) of the reference system according to eqn (2.17) and (2.18).

$$\chi_S = 1/2(IP_S + EA_S) \tag{2.17}$$

$$\eta_S = 1/2(IP_S - EA_S) \tag{2.18}$$

The IP_S and the EA_S of the reference system S are here defined as the reaction energies for the reactions shown in eqn (2.19) and (2.20):

$$S \xrightarrow{\text{IP}} S^{\oplus} + e^{\ominus} \tag{2.19}$$

$$S^{\ominus} \xrightarrow{\text{EA}} S + e^{\ominus} \tag{2.20}$$

and calculated according to eqn (2.21) and (2.22):

$$IP_S = E(N-1) - E(N) \tag{2.21}$$

$$EA_S = E(N) - E(N+1) \tag{2.22}$$

It is important to note that the definition of the electron affinity used here in eqn (2.20) is what is commonly referred to as the "mass spectrometric" definition starting from the one-electron reduced state of the reference system. An alternative "chemical" definition, reversing the direction of the reaction in eqn (2.20) is equally possible and leads to a change in sign in expressions (2.17) and (2.18). The consequences of the functional form of eqn (2.16) can best be illustrated using simple examples, such as fluorine and iodine atoms. Using experimentally known data for the ionization potentials ($IP(F) = +17.42$ eV, $IP(I) = +10.45$ eV) and electron affinities ($EA(F) = +3.40$ eV, $EA(I) = +3.06$ eV), we obtain the curves shown in Figure 2.13. In order to compare two systems with different absolute electron count, it is more convenient to plot the energies as a function of overall charge q. Both curves are thus anchored with $E_0 = 0.0$ eV at $q = 0$, that is, the energy E_{N0} in eqn (2.16) is set to zero for the free atoms. The energies of both systems at the $q = +1$ stage correspond to the ionization potentials of the respective atoms, while the energies at the $q = -1$ state correspond to the negative of the respective electron affinities. The absolute hardness of fluorine is calculated according to eqn (2.18) as $\eta(F) = +7.01$ eV, while that of iodine is $\eta(I) = +3.70$ eV. In graphical terms this can be seen in Figure 2.13 as a larger curvature of the parabolic curve for fluorine compared to that for iodine.

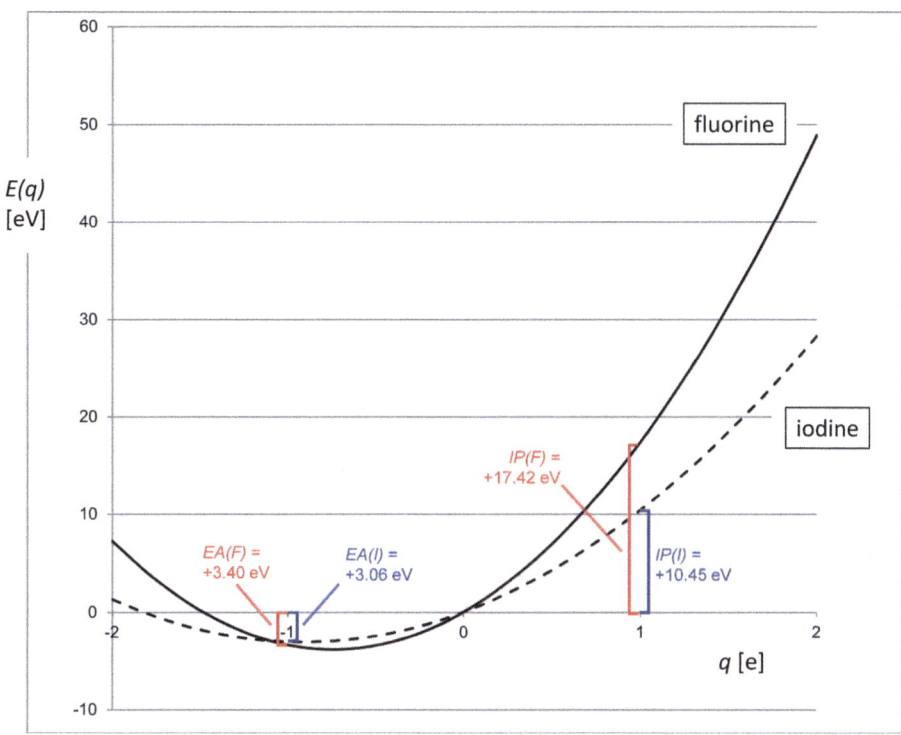

Figure 2.13 Energy of the fluorine and iodine atoms (in eV) as a function of atomic charge q.[5]

The determination of the hardness of molecular systems is often hampered by the lack of electron affinity data. For many (organic) molecules this is simply due to the fact that the reaction energy for the process defined in eqn (2.20) is negative, that is, the reactant anion is not bound with respect to electron loss. An extension of the HSAB principle has therefore been proposed to include results obtained in molecular orbital theory calculations. The basis of this extension is **Koopmans' theorem**, according to which the orbital energy of the highest occupied molecular orbital (HOMO) equates to the negative of the ionization potential, while the energy of the lowest unoccupied molecular orbital (LUMO) equates to the negative of the electron affinity according to eqn (2.23) and (2.24). How the frontier orbital energies relate to the absolute electronegativity χ_S and the absolute hardness η_S is shown in a pictorial way in Figure 2.14 for the example of hydrogen fluoride (HF) and hydrogen iodide (HI). The HI system is comparatively "soft" due to the comparatively high lying HOMO at -10.5 eV (and thus its ease of ionization), combined with a low lying LUMO at 0.0 eV that is at the verge of giving a positive electron affinity.

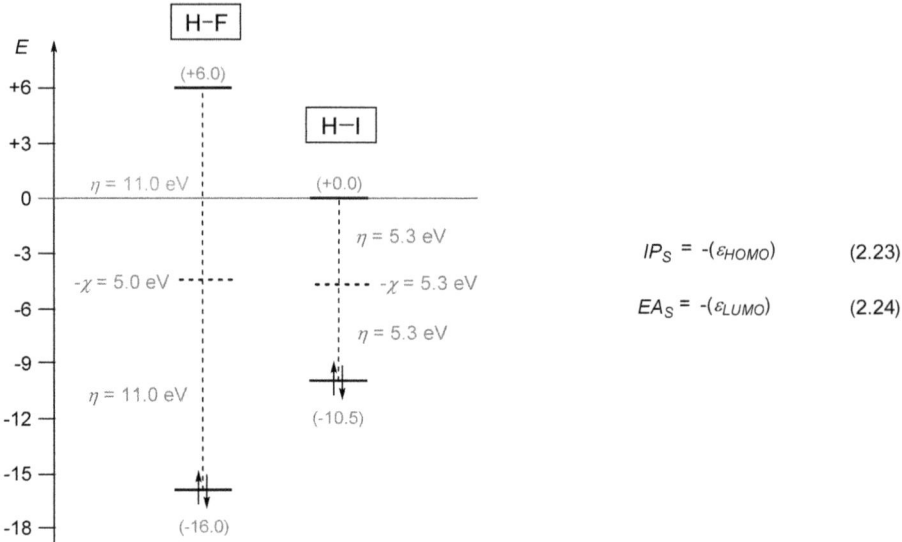

Figure 2.14 FMO energies, absolute electronegativity (χ) and absolute hardness (η) for HF and HI.[6]

The resulting small HOMO/LUMO gap of 10.5 eV gives, according to the definitions in eqn (2.17) and (2.18), electronegativity and hardness values of χ(HI) = 5.3 eV and η(HI) = 5.3 eV. HF, in contrast, has a much lower lying HOMO at −16.0 eV and is thus much harder to ionize. In addition, its LUMO energy is comparatively high at +6.0 eV, which leads to a rather large HOMO/LUMO gap of 22.0 eV and electronegativity/hardness values of χ(HF) = 5.0 eV and η(HF) = 11.0 eV. In a rather general sense, molecular hardness correlates with a large **HOMO/LUMO gap**.

With the MO-based definitions to hand one can go ahead and determine the electronegativity and hardness values of frequently used organic and inorganic molecules. Values for a small selection of systems are shown graphically in Figure 2.15. Visual inspection shows that oxidizing species are located more on the right and reducing species more on the left of the diagram, effectively following the calculated absolute electronegativity values. The hardest molecules, such as HF and BF_3, can be found at the top of the diagram and are small systems composed of first-row elements, such as fluorine and oxygen, while the softer molecules are located in the lower part of the diagram and contain either second-row or even heavier elements, or larger π-systems. That absolute electronegativity and hardness are not necessarily independent parameters is nicely illustrated by the halogen

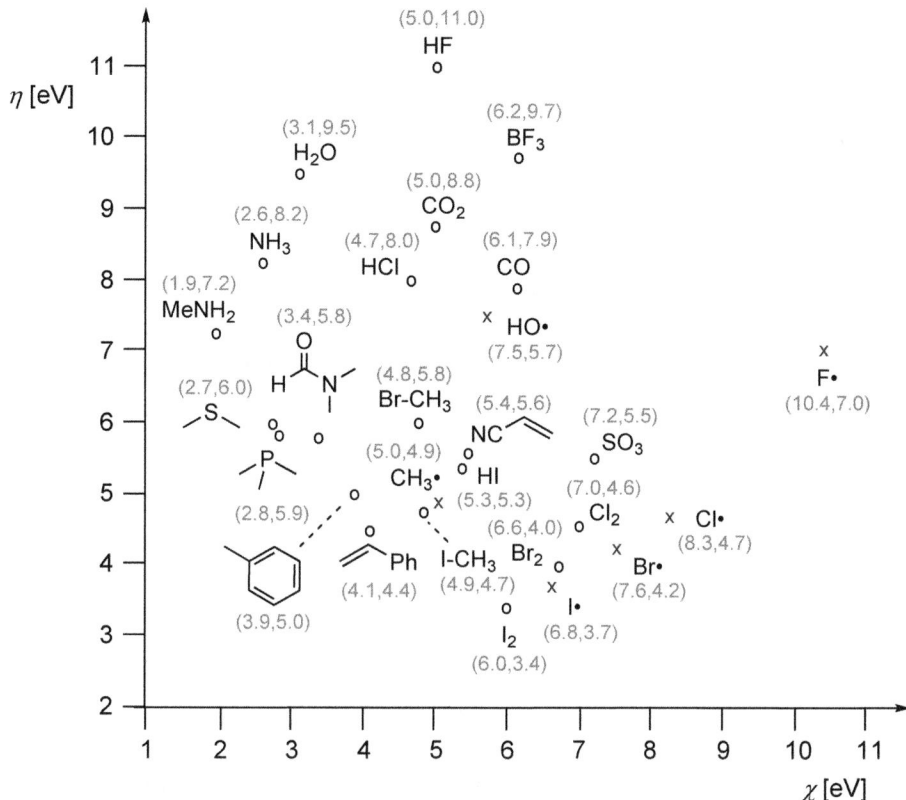

Figure 2.15 Absolute electronegativity (χ) and absolute hardness (η) for selected neutral systems. Radicals are designated by the "x" symbol.[6,7]

radical series, where increasing electronegativity is accompanied by systematically increasing hardness.

Applying the numerical definitions of χ and η to ions results in severe practical problems. This is easily seen for the proton, as one of the most important cations in molecular chemistry. Since the proton (H^+) has exactly zero electrons, its ionization energy is infinitely large, which implies infinitely large values for its electronegativity and hardness. But even for slightly larger cations, such as those of the alkali metals, the ionization energies are so large, that the resulting electronegativity and hardness values are far outside the diagram shown for neutral compounds in Figure 2.15. For many small anions, such as, for example, fluoride (F^-), there is no practical way to quantify their electron affinity, which effectively prevents the calculation of electronegativity and hardness values. Various proposals have been made to circumvent the problems described for cations and anions, but from a practical perspective it sometimes appears sufficient to define groups of comparatively "hard" and "soft" charged acids and bases without any attempt at quantification. From the systems shown in Figure 2.16 we see that the hard systems locate high positive or negative charge on comparatively little space and contain comparatively "light" elements. In contrast, the soft systems contain much heavier elements that are much easier to polarize.

A reaction commonly rationalized mechanistically on the basis of the HSAB reaction is the **Nef reaction** shown in Figure 2.17. Starting from primary or secondary nitroalkane substrates this reaction sequence involves initial deprotonation to the nitronate ion. Protonation of this latter species with strong acids can now occur at the O or the C terminus of the nitronate ion, a situation that is similar to the enolate regioselectivity discussed in Figure 2.12. Experimentally it is found that the "hard" O terminus of the nitronate ion reacts much faster than the "soft" C terminus, thus forming the aci-nitro isomer as a true intermediate of the reaction. This isomer is significantly less stable than the initially used nitroalkane and its formation is thus a consequence of kinetically controlled reactivity. From this it becomes immediately obvious that application of the HSAB principle is inappropriate simply due to its derivation from thermodynamic considerations, and analysis of the selective formation of the aci-nitro intermediate may

hard acids   **soft acids**

H^+ Li^+ Na^+ Mg^{2+} Al^{3+} Ag^+ Cu^+ Au^+ Pd^{2+}

hard bases   **soft bases**

HO^- RO^- F^- H^- I^- HS^- RS^- PhO^-

Figure 2.16   Selected groups of commonly used charged hard and soft acids and bases.[6]

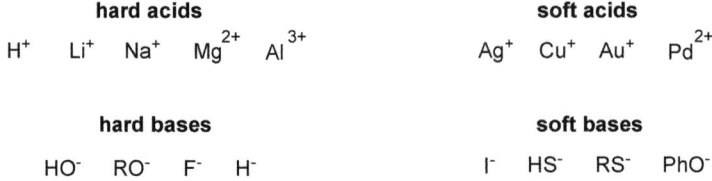

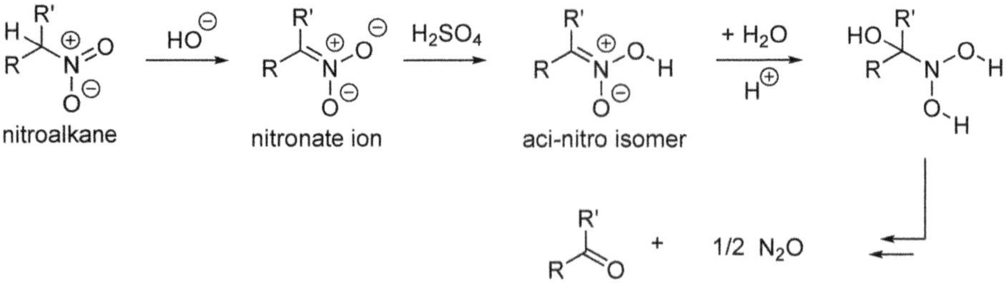

nitroalkane   nitronate ion   aci-nitro isomer

Figure 2.17   The Nef reaction for the conversion of nitroalkanes to aldehydes.

be better made on the basis of the Marcus idea of lower intrinsic barriers (again in close analogy to the reaction shown in Figure 2.12). In any case, the aci-nitro intermediate adds water to its formal C–N double bond faster than reverting back to the starting nitroalkane. The resulting hemiaminal-like second intermediate then hydrolyzes to the respective ketone and N_2O. As such the Nef reaction provides a systematic strategy for the conversion of primary or secondary nitroalkanes to the respective aldehydes or ketones.

2.5 Hammett Correlations

The basic hypothesis that reaction thermodynamics has some impact on reaction rates is also at the heart of the Hammett equation. L. P. Hammett originally quantified the effects of substituents X located at the *para*- or *meta*-position of benzoic acids on their acidity in aqueous solution at 25 °C. The equilibrium constant for the deprotonation reaction K_x as defined by eqn (2.25) has then been used, together with the equilibrium constant for unsubstituted benzoic acid K_H, to calculate the substituent constant σ_x (greek "sigma" symbol) according to eqn (2.26). This analysis excludes the effects of *ortho*-substituents, as these may influence K_x not only through electronic, but also through steric effects. The **Hammett substituent constants** collected in Table 2.1 show that electron-donating substituents in the *para*-position, such as –OH, reduced the acidity of benzoic acids and thus lead to negative σ_p values. In contrast, electron-withdrawing substituents, such as –CN, increase the acidity of benzoic acids, which is reflected in σ_p values with positive sign.

Table 2.1 Hammett substituent constants for selected substituents.[8]

Substituent	σ_p	σ_m	σ_p^+
–NMe$_2$	−0.83	−0.16	(−1.7)
–NH$_2$	−0.66	−0.16	(−1.3)
–OH	−0.37	+0.12	(−0.92)
–OMe	−0.27	+0.12	−0.78
–CH$_3$	−0.17	−0.07	−0.31
–H	0.0	0.0	0.0
–F	+0.06	+0.34	−0.07
–I	+0.18	+0.35	+0.14
–Cl	+0.23	+0.37	+0.11
–Br	+0.23	+0.39	+0.15
–CN	+0.66	+0.56	+0.66
–NO$_2$	+0.78	+0.71	+0.79

equilibrium
constant

$$X \underset{\substack{}}{\overset{}{\cdots}} \overset{O}{\underset{OH}{\bigcirc}} + H_2O \xrightleftharpoons[25\ °C]{K_x} X \overset{O}{\underset{O^{\ominus}}{\bigcirc}} + H_3O^{\oplus} \tag{2.25}$$

substituent
constant

$$\log\left(\frac{K_x}{K_H}\right) = \sigma_x \tag{2.26}$$

How the rate of a reaction responds to changes in the substitution pattern can be analyzed using the Hammett equation (2.27), where k_x corresponds to the rate constants for reaction of the substrate carrying substituent X and k_H is the rate constant for the unsubstituted reference system.

rate
constant   substituent
constant

$$\log\left(\frac{k_x}{k_H}\right) = \sigma\,\rho \tag{2.27}$$

reaction
constant

Gibbs free energies
of activation

$$\frac{\Delta G_H^\ddagger - \Delta G_X^\ddagger}{2.303\ RT} = \sigma\,\rho \tag{2.28}$$

How susceptible the reaction under study is to changes in the substitution pattern (and thus changes in the substituent constants σ) is indicated by the Hammett reaction constant or **Hammett ρ value** (greek "rho" symbol). When combining eqn (2.27) with the Eyring equation, one can see that the left-hand side of eqn (2.27) actually corresponds to the difference in the Gibbs free energies of activation for reaction of the unsubstituted and the substituted systems, as shown in eqn (2.28). The Hammett equation is therefore also called a **linear free energy relationship**. How the Hammett equation is used in practice is shown in Figure 2.18 for the base-induced hydrolysis of ethyl benzoates in ethanol/water (85:15) as an example. The reaction shows quite large substituent effects and is approximately four orders of magnitude faster for X = p-NO$_2$ compared with X = p-NH$_2$. Plotting

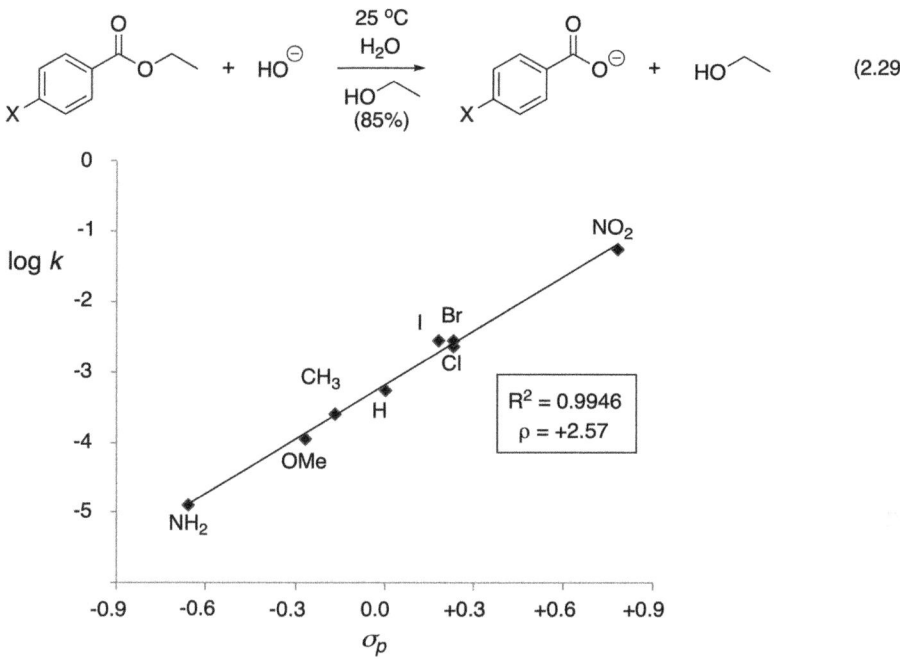

Figure 2.18 Hammett plot of reaction rate constants for the base-induced hydrolysis of ethyl benzoates.[9]

the logarithmic rate constants against the σ_p values shown in Table 2.1 yields the Hammett plot shown in Figure 2.18. Linear regression analysis shows a good correlation coefficient ($R^2 = 0.9946$) and yields a Hammett reaction constant of $\rho = +2.57$. In pictorial terms the Hammett ρ value is identical to the slope of the correlation line shown in Figure 2.18, and its large positive value found here is commonly interpreted as an increase in negative charge in the rate-limiting transition state compared to the ground state.

The quality of Hammett plots employing σ_p parameters often deteriorates in reactions where a positive charge forms either inside the aromatic ring system or in a benzylic position. A case in point is the hydrolysis of cumyl chloride shown in Figure 2.19, which is generally assumed to proceed through an S_N1-type mechanism. In the cumyl cation intermediate formed in this reaction, direct resonant interactions are possible between the reaction center and substituents in the *para-* (but not in the *meta-*) position. This led Brown and Okamoto to define a new set of substituent parameters termed σ_p^+ parameters. In a first step, a reaction constant of $\rho = -4.54$ was determined for

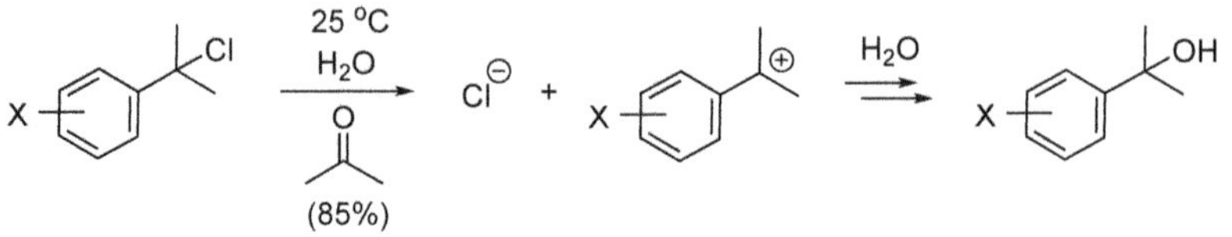

$$\log\left(\frac{k_X}{k_H}\right) = \sigma_p^+ \rho \qquad (2.31)$$

Figure 2.19 Hydrolysis of cumyl chloride substrates in acetone/water mixtures.[10]

reactions involving only *meta*-substituted cumyl chloride substrates using the unchanged σ_m parameters. This is in full agreement with the assumption of an increase of positive charge in the rate-limiting transition state. With this reaction constant in hand the reaction rates for *para*-substituted cumyl chloride substrates were then used to determine the new σ_p^+ parameters listed in Table 2.1. These are not only practical for describing reactions with transient benzyl cation intermediates, but also for aromatic electrophilic substitution reactions.

Hammett equations are often used to analyze the mechanism of a reaction. An example is the reaction of substituted phenyldiazomethanes with alkenes shown in Figure 2.20a. Phenyldiazomethanes are known to react with electron-poor alkenes in a Huisgen (3 + 2) cycloaddition to yield 5-arylpyrazolines as the primary reaction products. In order to characterize the polar or charge-separating character of this transformation, reaction rates for four *para*-substituted phenyldiazomethanes with bis(phenylsulfonyl)ethylene (**A**) as a model electron-poor alkene have been compared to those for reaction with substituted benzyhydrylium tetrafluoroborate **B**. This latter transformation is assumed to proceed through initial (and rate-limiting) attack of the benzyhydryl cation at the diazomethane carbon atom, followed by N_2 elimination and rearrangement steps. This mechanism is in full agreement with a reaction constant of $\rho = -2.34$ found in the Hammett analysis (Figure 2.20b). It is therefore quite interesting to see an equally negative reaction constant of $\rho = -2.42$ for reaction of phenyldiazomethanes with bis(phenylsulfonyl)ethylene. This can either be interpreted as a stepwise cycloaddition reaction involving initial electrophilic attack of the less substituted alkene terminus at the diazomethane

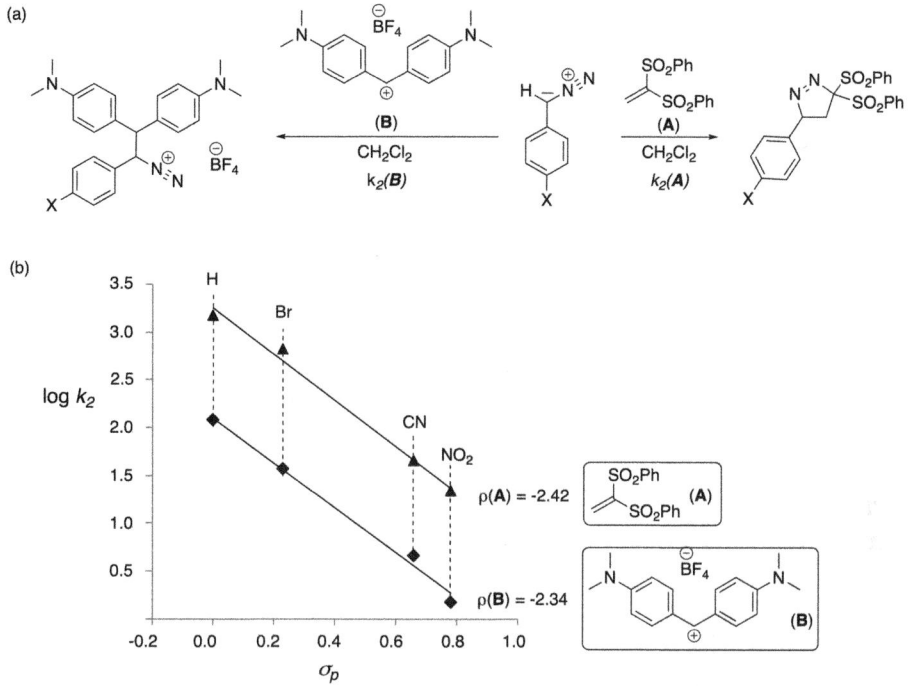

Figure 2.20   (a) Reaction of phenyldiazomethanes with bis(phenylsulfonyl) ethylene (**A**) and benzhydrylium cation salt **B**. (b) Hammett analysis of the reaction of phenyldiazomethanes with **A** and **B** (rate data from ref. 11).

carbon atom, or as a concerted but quite asynchronous cycloaddition process.

2.6   The Mayr–Patz Equation

The Mayr–Patz equation (2.32) provides a convenient way for expressing the rates of bimolecular electrophile/nucleophile reactions in a systematic way. The second-order rate constant k at $T = 20\ °C$ is expressed here as a function of the electrophilicity parameter E, the nucleophilicity parameter N and the nucleophile-specific slope parameter s_N. Both N and s_N are solvent dependent. All three parameters have been determined from experimental rate measurements for bimolecular reactions where only one new bond is formed between the reactants and where the overall charge of reactants and products remains constant.

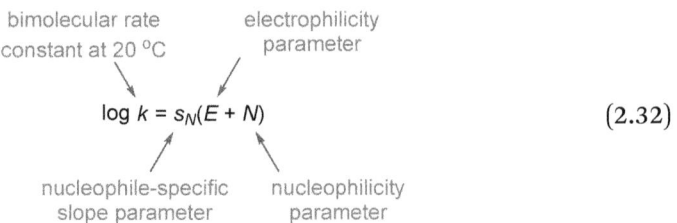

$$\log k = s_N(E + N) \tag{2.32}$$

A typical example is shown in Figure 2.21, where selected benzhydryl cations are used to quantify the reactivity of enamines **A** and **B**, and of silylenolether **C** in CH_2Cl_2 solution. The benzyhydryl cations are employed here as the respective tetrafluoroborate salts and cover a comparatively large range of reactivities. At the high reactivity end we find the $(ani)_2CH^+$ cation characterized by $E = 0.0$ and at the low reactivity end the $(lil)_2CH^+$ cation with $E = -10.04$ (see Figure 2.22 for the full cation structures). Keeping in mind that E parameters express reaction rates on a logarithmic scale, this represents a reactivity difference of ten orders of magnitude. Despite these large differences, the reaction centers of all benzhydryl cations used here have practically identical structures, and therefore variations in reactivity reflect differences in their underlying electronic structures. The most reactive nucleophile studied here is enamine **A**, characterized by $N = +13.36$ and $s_N = 0.81$. The oxygen atom present in the morpholino-substituted enamine **B** reduces its nucleophilic reactivity by almost two orders of magnitude and leads to $N = +11.40$ and $s_N = 0.83$. Enolether **C**, finally, is much less reactive than both enamines, with $N = +5.21$ and $s_N = 1.00$. In graphical terms, the N parameter represents the crossing point of the regression line for a given nucleophile with the horizontal line where $\log k = 0$ (which is also where $N = -E$), while the s_N parameter equates to the slope of the regression line.

Effectively following the same strategy as outlined above, Mayr and coworkers have derived reactivity parameters for a wide variety of nucleophiles and electrophiles, a small selection of which is shown in Figure 2.22. Synthetically useful reactions between electrophiles and nucleophiles can be expected for reactions where $E + N > -5$.

The validity of eqn (2.32) can be tested with electrophile/nucleophile combinations not employed to derive the respective reactivity parameters. Sulfonium ylide **A** shown in Figure 2.23, for example, is characterized by N/s_N values of $+11.95/0.76$ and is thus expected to show synthetically useful reactivity with electrophiles with $E > -17$. Why reactions of **A** are not successful under organocatalytic conditions

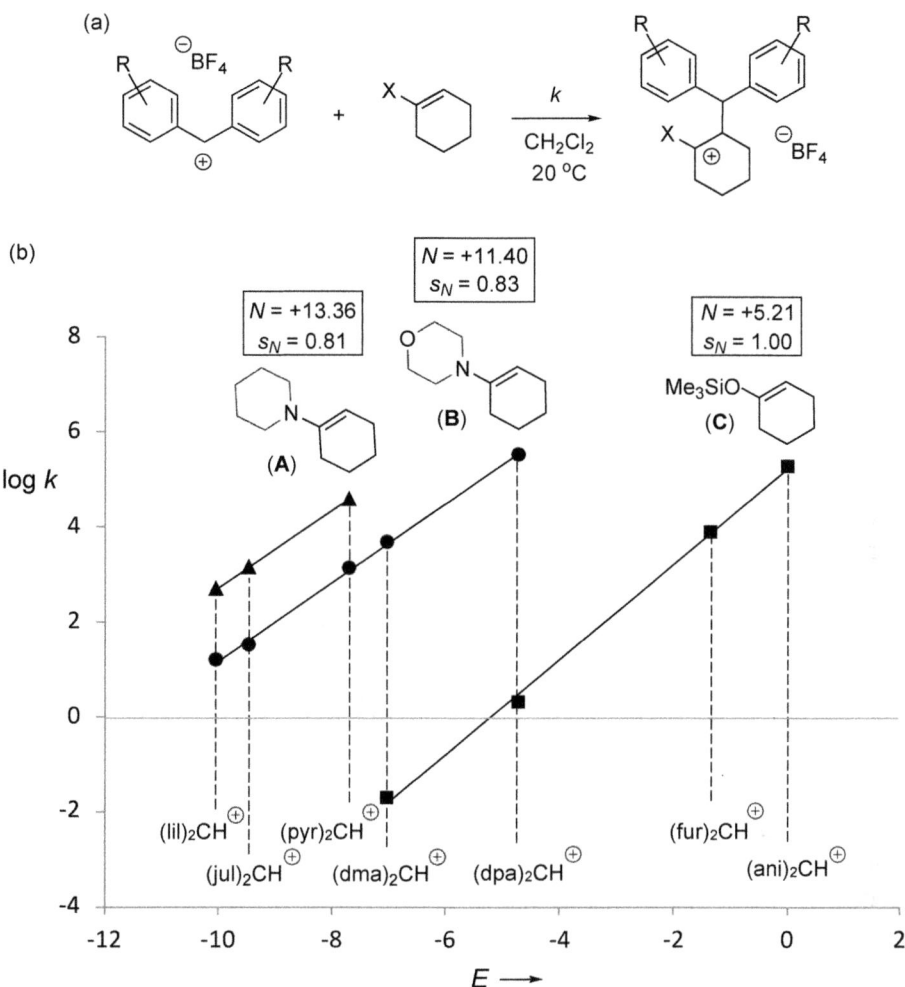

Figure 2.21 (a) Addition reaction of benzhydrylium cations with substituted cyclohexene nucleophiles. (b) Reaction rates k for addition of 1-(N-piperidino)cyclohexene (**A**), (N-morpholino)cyclohexene (**B**) and 1-(trimethylsilyloxy)cyclohexene (**C**) to selected benzhydryl cations plotted against the respective E parameters (rate data from ref. 12).

with electrophile **B** ($E = -7.37$) is thus not immediately apparent. Inserting the above reactivity parameters into the Mayr–Patz equation predicts a rate constant of $k = 3.0 \times 10^2 \ M^{-1} \ s^{-1}$. The reaction half-lives of ordinary second-order reactions depend inversely on the initial concentrations c_0 and the rate constant (that is, $t_{1/2} = 1/(k \cdot c_0)$), and assuming initial concentrations of 1 M we arrive at reaction half-lives

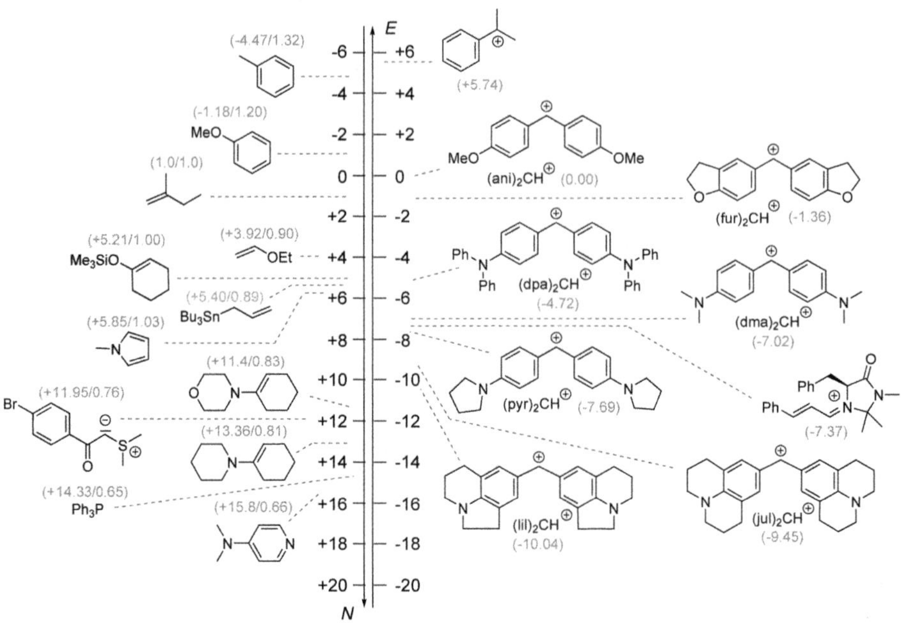

Figure 2.22 Reactivity parameters for selected electrophiles and nucleophiles. E parameters are given in parentheses below all structures of electrophiles, while N/s_N parameters are given in parentheses below all structures of nucleophiles (for an up-to-date list of parameters see www.cup.lmu.de/oc/mayr/reaktionsdatenbank2/).[12–14]

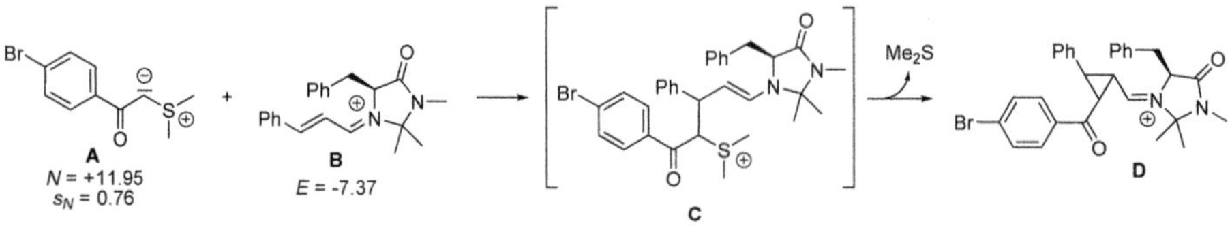

Figure 2.23 Reaction of sulfonium ylide **A** with iminium ion **B**.[15]

of approximately $t_{1/2} = 3 \times 10^{-3}$ s for the reaction of **A** and **B**. Independent synthesis of electrophile **B** in a non-catalytic fashion and reaction with **A** does indeed proceed readily and yields the zwitterion **C** as the initial adduct, from which the elimination of dimethyl sulfide yields cyclopropane **D** as the final product. The lack of reactivity of ylide **A** under organocatalytic conditions is therefore not due to its sluggish addition to electrophile **B**, but likely derives from its undesired reaction with a proton source.

2.7 The Klopman–Salem Equation and FMO Theory

Perturbation MO theory has independently been employed by G. Klopman and L. Salem to derive equations describing the energy change involved in the reactive encounter of two molecular systems. The currently used **Klopman–Salem equation** (Figure 2.24) combines interactions resulting from specific sets of orbitals with a classical expression for electrostatic interactions. Taking the interaction of buta-1,3-diene (labeled **R**) and acrolein (labeled **S**) in a Diels–Alder transition state as an example, the first term describes closed-shell repulsion interactions between the occupied orbitals of the reactants **R** and **S**. The magnitude of this interaction depends on the atomic orbital populations in **R** and **S** (q_a and q_b), the resonance integrals β_{ab} and the overlap integrals S_{ab}. The second term describes classical electrostatic interactions between the reactants **R** and **S**, and can be repulsive or attractive in nature. It depends on the size of the atomic charges Q_k in **R** and Q_l in **S**, the distance R_{kl} between the interacting atomic sites and the local dielectric constant ε. The third term describes the interaction of occupied orbitals in **R** with unoccupied (or "virtual") orbitals in **S**, and *vice versa*. It depends on the molecular orbital coefficients c_{ra} in **R** and c_{sb} in **S**, as well as the resonance integral β_{ab}. Most importantly (and as we have seen already in Chapter 1, Section 1.2 on qualitative

$$\Delta E = \underbrace{-\sum_{ab}(q_a + q_b)\beta_{ab}S_{ab}}_{\text{1. term}} + \underbrace{\sum_{k<l}\frac{Q_kQ_l}{\varepsilon R_{kl}}}_{\text{2. term}} + \underbrace{\sum_r^{\text{occ}}\sum_s^{\text{unocc}} - \sum_s^{\text{occ}}\sum_r^{\text{unocc}} \frac{2\left(\sum_{ab}c_{ra}c_{sb}\beta_{ab}\right)^2}{(E_r - E_s)}}_{\text{3. term}} \tag{2.33}$$

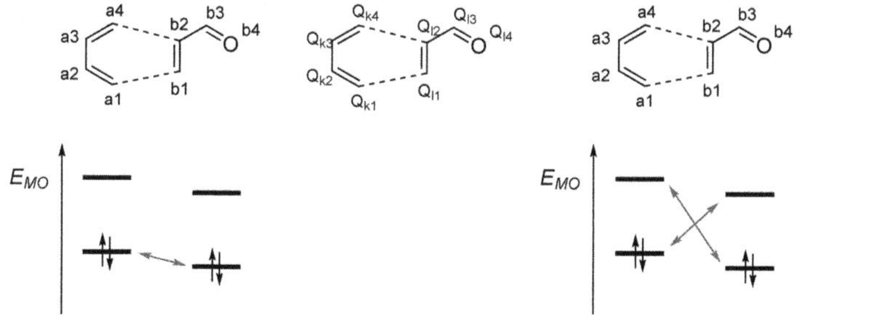

Figure 2.24 Klopman–Salem equation (2.33) together with schematic representations of the three interaction energy terms.[16]

MO theory) the interaction energy depends inversely on the energy difference between the interacting orbitals in **R** and **S**, and will thus be particularly large for the interaction of the HOMO in one of the reactants and the lowest unoccupied orbital (LUMO) in the other. An important general property of eqn (2.33) is that, aside from some integral values, the interaction energy is expressed in terms of the properties of the separate, non-interacting reactants **R** and **S**. The orbital energies E_r as well as the MO coefficients c_{ra} in the third term are therefore those of the unperturbed reactant **R**, despite the fact that the overall energy expression in eqn (2.33) attempts to relate the energy of the transition state to that of the ground state.

When applying eqn (2.33) to reactivity problems in organic chemistry, it is often assumed that the repulsive interactions described by the first term are more or less comparable for competing reaction channels, leading, for example, to different regioisomers. An example illustrating this situation is the electrophilic aromatic substitution of pyrrole in its reaction with acetyl nitrate. As shown in Figure 2.25a, this reaction yields 2-nitropyrrole as the exclusive product isolated in 55% yield, most likely through *in situ* formation of nitronium ion (NO_2^+), followed by attack at the pyrrole C2 position and deprotonation to the final product. Drawing out the more

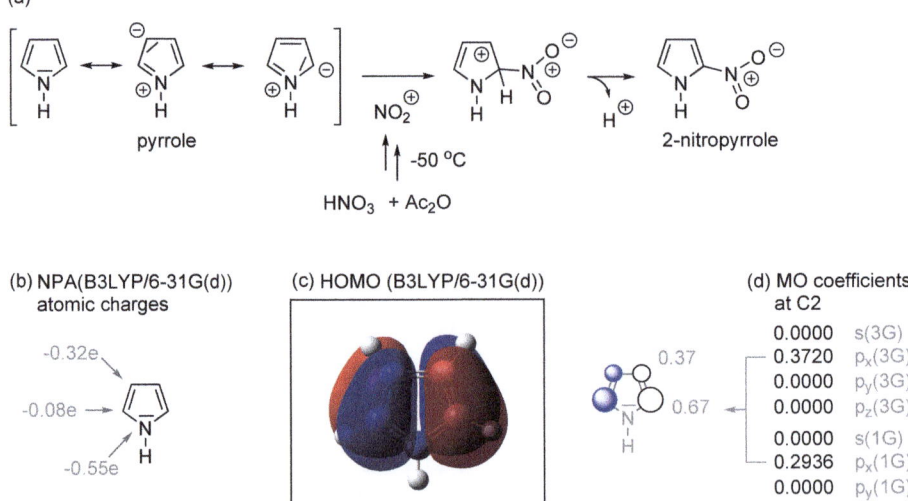

Figure 2.25 (a) Nitration of pyrrole with HNO_3/acetic anhydride. (b) Charge distribution in pyrrole. (c) HOMO of pyrrole. (d) MO coefficients for the pyrrole HOMO.[17]

relevant Lewis structures for pyrrole, as shown in Figure 2.25a, does not immediately indicate a preference for addition to either the C2 or the C3 position.

Whether the regioselectivity of this process is dominated by electrostatic effects (that is, the second term of the Klopman–Salem equation) can be analyzed by examination of the partial charges calculated quantum mechanically with the natural population analysis (NPA) approach (here at the B3LYP/6-31G(d) level of theory). The most negative partial charge (in units of the elemental electron charge e) is found on the ring nitrogen atom (−0.55), followed by the C3 position (−0.32) and the C2 position (−0.08). The hydrogen atoms not shown explicitly in Figure 2.25b all carry positive partial charges. The numerical values obtained with other quantum mechanical methods or population analysis schemes differ a little from those shown in Figure 2.25b, but the charge differences between the C2 and C3 positions remain the same in all cases. This charge distribution is obviously not in line with preferential attack of the positively charged nitronium ion at the C2 position, as we find much less (if any) negative charge here than on C3. The third term of the Klopman–Salem equation is, in this case, dominated by the HOMO(pyrrole)/LUMO(nitronium ion) interaction, and the MO coefficients of the pyrrole HOMO are therefore decisive here. Visual inspection of the three-dimensional (3D) plot of the HOMO calculated with the B3LYP/6-31G(d) method clearly shows a larger extent at C2 than at C3 (Figure 2.25c). When analyzing the underlying HOMO coefficients at the C2 and C3 positions, we encounter a technical problem resulting from the nature of the basis sets used in contemporary quantum mechanical methods. At the stage of its development, the Klopman–Salem equation (2.33) was applied using MO coefficients derived from rather economical quantum chemical methods, such as the Hückel MO theory. These methods employ single p-type (atomic) basis functions for the description of p-type valence space orbitals. The more sophisticated theoretical methods employed currently use so-called "split valence" basis functions, where two different p-type basis functions are employed instead. The actual characteristics of these valence space basis functions are encoded in the basis set name, and the "31" in the 6-31G(d) basis set indicates that the valence space s- and p-type orbitals are described by combination of a more compact "inner" basis function constructed from three Gaussian functions and a second "outer" basis function represented by a single Gaussian. The actual mixture between these two components is determined at the stage of optimizing the molecular orbitals, and the result of this process is shown for the HOMO of pyrrole at the C2 position in Figure 2.25d. The list of coefficients has only two significant entries here, with 0.3720 for the p_x(3G) and 0.2936 for the p_x(1G) basis functions. The fact that all the other coefficients are 0.00 is simply due all the atoms of pyrrole being located in the yz plane of the coordinate system. Summing up the two p_x coefficients we obtain an "effective" coefficient of 0.67 for the C2 position in the pyrrole HOMO, which compares to the C3 coefficient of 0.37 obtained in the same way. The fact that electrophilic attack occurs preferentially at the pyrrole C2 position can therefore be rationalized satisfactorily by the HOMO structure. This is also found for electrophilic substitution reactions in many other carbo- or heterocyclic aromatic systems, and the calculation of MO coefficients for the highest occupied orbitals in these systems represents a valuable tool for the prediction of regioselectivities.

While the electrostatic interactions described by the second term of the Klopman–Salem equation are highly relevant in guiding the reactivity of charged or dipolar species, there is the general feeling that the third term dominates in reactions of uncharged molecules. Analyzing reactivity patterns exclusively in the framework of the highest occupied and the lowest unoccupied orbitals is usually referred to as **frontier molecular orbital (FMO)** theory as developed by K. Fukui. A typical area of application is radical chemistry, where neutral C-centered radicals are often found to behave as nucleophiles in their addition to alkenes. For example, the room–temperature reaction of *tert*-butyl radical with ethylene is more than three orders of magnitude slower than the reaction with acrylonitrile (Figure 2.26). The energy gained through interaction of the singly occupied orbital (SOMO) of the *tert*-butyl radical with the alkene LUMO is larger than that through interaction with the alkene HOMO, even though this might not immediately be apparent from relative orbital energies. Comparing frontier orbital energy levels in ethylene and acrylonitrile, we can readily see that the acrylonitrile LUMO is located at much lower energies compared to the ethylene LUMO. Interaction between the radical SOMO and the acrylonitrile LUMO therefore involves the smaller energy difference and leads to larger energy stabilization compared to ethylene, which is in full agreement with the experimentally observed faster addition rate for acrylonitrile. Visual inspection of the acrylonitrile LUMO also indicates a strong polarization towards the unsubstituted alkene terminus, and so the preference for radical addition at this position may also be rationalized by the picture of dominating SOMO(radical)/LUMO(alkene) interactions in the transition state of the addition step. It should be reiterated at this point that it is not the

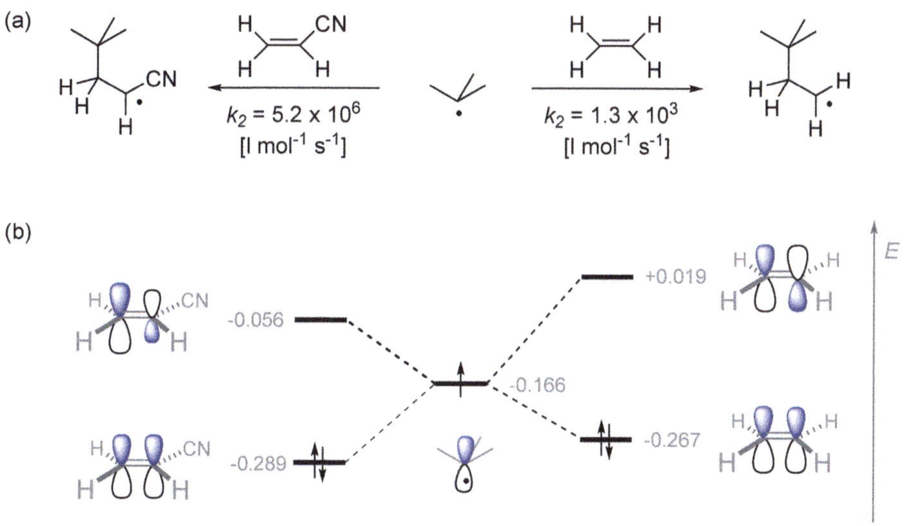

Figure 2.26 (a) Addition reaction of *tert*-butyl radical with ethylene and acrylonitrile and (b) selected molecular orbital energy levels (in Hartree, calculated at B3LYP/6-31G(d) level).[18]

SOMO or LUMO of the transition state that is analyzed here, but the frontier molecular orbitals of the unchanged and separate reactants.

FMO theory is also invoked to rationalize general trends in organic reactivity. For a variety of nucleophiles, for example, it is generally observed that addition rates to carbonyl compounds depend dramatically on the substituents attached to the C–O double bond. Rates are commonly found to decrease in the series aldehydes > ketones > esters > amides. The leading FMO interaction in nucleophilic addition reactions to these compounds is the HOMO(nucleophile)/LUMO(electrophile) interaction, which implies that lower LUMO levels will lead to larger interaction energies and thus faster reaction rates. As shown in Figure 2.27, the computed LUMO energies of formaldehyde, acetaldehyde, acetone, methyl acetate and *N*-methyl acetamide follow exactly the order expected from their relative reactivities towards nucleophiles.

How different the relative rates for nucleophilic addition reactions to aldehydes and ketones can be is nicely demonstrated by the relative reduction rates for benzaldehyde and acetophenone with $NaBH_4$ in isopropanol at 0 °C (Figure 2.28). The reduction of benzaldehyde is 392 times faster than that of acetophenone, despite the seemingly small structural and electronic differences between these two substrates. From a synthetic point of view this also implies that aldehydes can be reduced selectively even in the presence of a competing ketone.

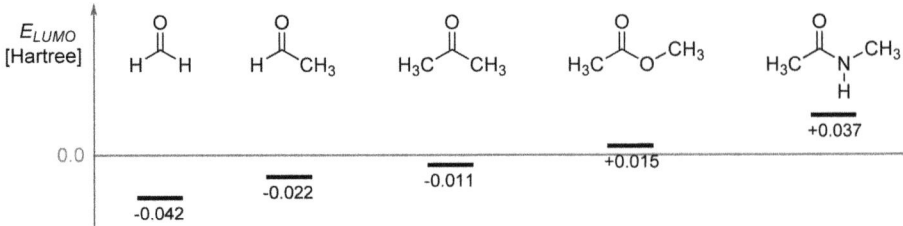

Figure 2.27 LUMO energy levels for selected carbonyl compounds (in Hartree, calculated at B3LYP/6-31G(d) level).

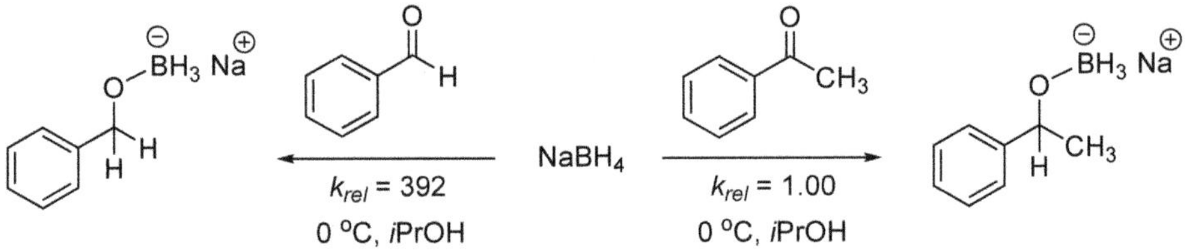

Figure 2.28 Reduction rates for benzaldehyde and acetophenone with NaBH$_4$.[19]

2.8 Solvent Effects

Most reactions in organic chemistry are performed in the liquid phase for practical reasons. The "organic" solvents used to provide a common (liquid) phase for all reaction partners serve multiple purposes and also influence the reaction kinetics and thermodynamics. The rationalization of these solvent effects is often based on the analysis of intermolecular (non-covalent) interactions of the dissolved substrates with single solvent molecules. Therefore, in the following sections we will first take a closer look at the most common non-covalent interaction types. Then, in a second step, we will see how bulk solvent effects can be analyzed and quantified.

2.8.1 Non-covalent Interactions

Before looking at actual numerical values for selected non-covalent interactions we recall from Section 2.1 that bond energies appear in the literature with different definitions. When a bond energy relates to the vibrationless potential energy at 0 K, it is often referred to as "D_e". Additional consideration of zero-point vibrational energies leads to a systematic reduction of bond energies, which are then referred to as "D_0". Comparison with other thermochemical data is most practical when using standard-state bond enthalpies at 298.15 K, which may be abbreviated in a number of different ways, such as, for example, "BDE" (bond dissociation energy). In the following we will concentrate on BDE values for selected non-covalent interactions simply because this will facilitate direct comparison with the covalent bond energies discussed in Chapter 1, Section 1.4.1.

Interactions between anions and cations count as being amongst the strongest non-covalent interactions in molecular systems. An often cited example involves the interaction of sodium cation (Na^+) with chloride anion (Cl^-) to form a single molecular "NaCl" entity in the gas phase. The gas-phase bond dissociation energy of this adduct for dissociation into the constituent ions amounts to BDE(NaCl) = +412 kJ mol^{-1}, which puts this non-covalent interaction directly into the range of typical (covalent) C–H bond energies discussed before! The structure of this ion pair is quite compact, with a sodium-chloride distance of r(NaCl) = 236 pm, which is actually slightly shorter compared with the solid-state NaCl (table salt) structure (Figure 2.29a).

Interactions between ions of opposite or like charge are commonly discussed in terms of classical **electrostatics**, as described by **Coulomb's law** (eqn (2.34)). The particular form chosen here expresses the gas-phase electrostatic potential energy of a system composed of point charges q_1 and q_2 as a function of the center-to-center distance r_{12}. The constant ε_0 appearing in the denominator is the vacuum permittivity. It is easy to see that the electrostatic interaction energy is negative (that is, attractive) for oppositely charged ions, while it is positive (that is, repulsive) for the interaction of ions of like charge. From the

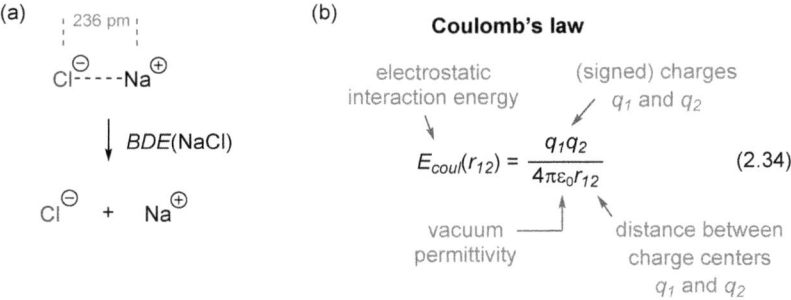

Figure 2.29 (a) Dissociation of the sodium chloride ion pair into its constituent ions and (b) Coulomb's law.[20,21]

functional form of eqn (2.34) it is also apparent that the electrostatic interaction energy is linearly proportional to the inverse distance r_{12}, with shorter distances translating into stronger interactions.

The fact that Coulomb's law, as described by eqn (2.34), is not sufficient to describe the interactions of atomic or molecular ions follows from the interatomic distance of 236 pm shown in Figure 2.29a for the sodium chloride ion pair. As the electrostatic energy will further increase on reducing the interatomic distance, there must be other forces in place that prevent the atoms from approaching each other any further. The answer can be found at the weak end of non-covalent interactions in, for example, the argon dimer, where two atoms of the rare gas argon interact to form a dimer located in an exceedingly shallow potential energy well.

The structure of this dimer is quite "loose" with an atom-to-atom distance of $r(\text{Ar–Ar}) = 377$ pm and an interaction energy of $D_e(\text{Ar–Ar}) = 1.2$ kJ mol^{-1}. A quantitative description of the interaction of neutral closed-shell systems, such as the argon dimer, where electrostatics are obviously not involved, can be approached using the **Lennard-Jones potential**, as expressed in eqn (2.35) shown in Figure 2.30. The potential energy of the system E_{LJ} here results from a repulsive and an attractive part. The repulsive part scales with the twelfth power of the ratio of the distance parameter σ and the interatomic distance r_{12}, while the attractive part scales with the sixth power. The distance parameter σ is defined as the distance where E_{LJ} equates to zero, the actual minimum of the potential energy well being located at $r_{min} = 2^{(1/6)}\sigma$. The sum of the attractive and repulsive terms is multiplied by the well-depth parameter ε (which bears no relation to the vacuum permittivity shown in Coulomb's law in eqn (2.34) which uses the same greek

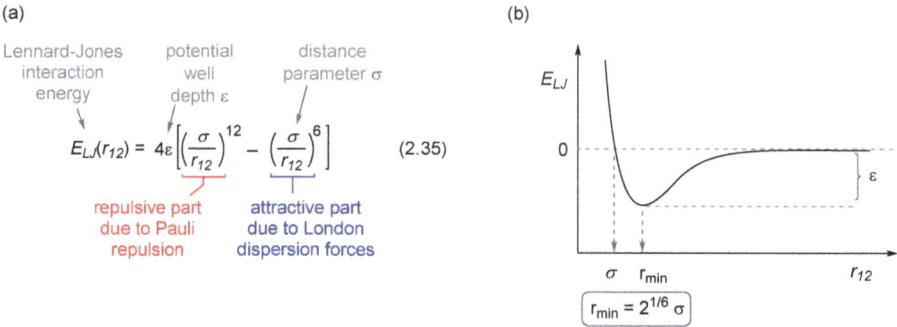

Figure 2.30   (a) Eqn (2.35) describing the Lennard-Jones potential and (b) a pictorial representation of eqn (2.35).[22]

symbol). Due to the large exponents for the repulsive and attractive terms in eqn (2.35), the Lennard-Jones interaction energy falls off very quickly with increasing distance r_{12}. The repulsive term describes the **Pauli repulsion** between occupied orbitals of the interacting molecular systems, while the attractive part is due to **London dispersion** energies resulting from induced dipole-induced dipole interactions. These two forces are also active in the example of the sodium chloride ion pair discussed previously, and it is the rapidly rising Pauli repulsion that keeps the atoms from approaching closer than the equilibrium distance shown in Figure 2.29a. The attractive London dispersion contribution adds little to the binding energy of the sodium chloride ion pair, simply because the electrostatic attraction is overwhelmingly large. However, due to the pairwise nature of these interactions, quite significant contributions can accumulate in the interactions of larger molecular systems.

The combined action of the three principal forces that we have seen at work in the sodium chloride and argon dimer examples (electrostatics, Pauli repulsion and London dispersion) will be sufficient to discuss most of the non-covalent interactions present in solutions of molecular systems. **Hydrogen-bonding** interactions are among the most common interaction types. A prototypical example is the hydrogen bond between two water molecules, where one acts as a hydrogen bond donor and the other as the acceptor. Dissociation of the water dimer into two separate water molecules is endothermic by $BDE(H_2O-HOH) = +15.7$ kJ mol^{-1} and mainly derives from electrostatic interactions between the partially positively charged hydrogen atom in one water molecule with the partially negatively charged oxygen atom in the other. In structural terms the hydrogen bond angle (that is, the O–H$\cdots$O) deviates only minimally from linearity, and the oxygen atoms of the two water molecules are positioned at a distance of approximately 295 pm. Together with O–H bond distances in water of around 95 pm, this leaves us with an actual hydrogen bond distance (that is, H$\cdots$O) of around 200 pm (Figure 2.31).

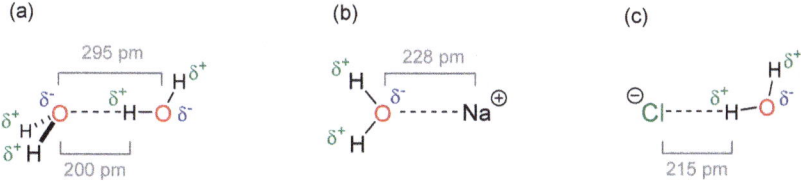

Figure 2.31 The water dimer (a) together with water complexes of a sodium cation (b) and chloride anion (c).[23–25]

Electrostatic interactions dominate the interaction of polar solvent molecules, such as water, with positively or negatively charged ions. We again pick the sodium cation here as an example for singly positively charged ions, whose water complex is characterized by an O–Na distance of 228 pm and a bond energy of $BDE(H_2O - Na^+)$ = +104.6 kJ mol^{-1}. The water complex of chloride, as a typical monovalent anion, is characterized by a single hydrogen bond of 215 pm length and a $BDE(Cl^- - H_2O)$ value = +60.2 kJ mol^{-1}. This latter value nicely illustrates that particularly strong hydrogen bonds are formed when negatively charged acceptors are involved. This is similarly true for the combination of positively charged hydrogen bond donors with neutral acceptors.

Electrostatic interactions also provide a qualitative guide to understand the interactions of charged species with organic solvents, which are considered to be apolar in nature. An important example here is benzene, not so much because of its use as a solvent in organic chemistry, but as the prototype for interactions of cations with π-systems. As shown schematically in Figure 2.32a, the benzene C–H bonds are polarized such that partial negative charges accumulate at the carbon atoms. In the benzene/sodium cation complex, electrostatic interactions are optimized through positioning the sodium cation directly at

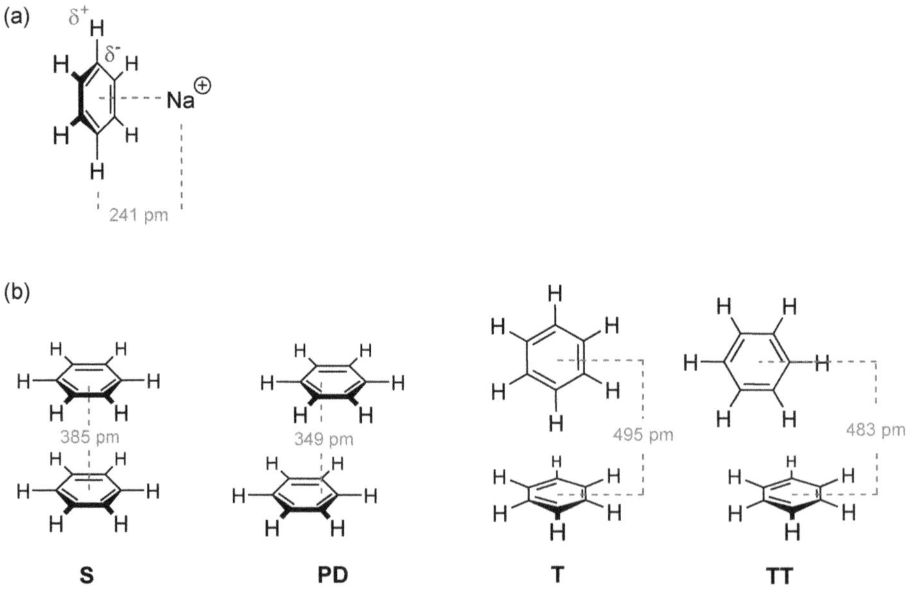

Figure 2.32 Benzene complexes with (a) sodium cation and (b) a second benzene molecule.[26-28]

the center of one of the benzene π-faces. The binding energy between the two components is quite strong at $BDE(C_6H_6-Na^+) = +94.3$ kJ mol^{-1} and thus only marginally smaller compared with the sodium cation/water complex. **Cation/π-complexes**, such as the example shown in Figure 2.32, occur widely in protein structures, where the protonated (and thus positively charged) side chains of arginine and lysine interact with the aromatic residues in phenylalanine, tryptophan or tyrosine.

The benzene dimer is a typical system, where two π-systems interact through a combination of electrostatic and London dispersion forces. For the sake of simplicity this system is often drawn in the sandwich dimer structure **S** shown in Figure 2.32b, but this structure does not represent a true minimum on the potential energy surface. This is easily understood on the basis of simple electrostatics, as this will position the partially negatively charged carbon atoms in one benzene ring right on top of the equally negatively charged carbon atoms in the other. From various experimental and theoretical studies it now seems to be established that the parallel-displaced (**PD**), the T-shaped (**T**) and the twisted T-shaped (**TT**) structures are more representative and all similarly stable, with dissociation energies of $D_e(\mathbf{PD}) = 11.3$ kJ mol^{-1}, $D_e(\mathbf{T}) = 11.3$ kJ mol^{-1} and $D_e(\mathbf{TT}) = 11.6$ kJ mol^{-1} relative to two separate benzene monomers. In structural terms the **PD** conformation is characterized by a center-to-center distance of the two benzene rings of 349 pm, a value often assumed to be typical for π-systems in a **stacking conformation**. In the **T** and **TT** dimers the C–H bonds in one benzene monomer point towards the ring face of the other, the center-to-center distances now being around 490 pm. A short comment on nomenclature seems appropriate here. The term "stacking interactions" repeatedly found in the literature should be avoided as it suggests the presence of specific interaction types favoring stacking conformations, such as the benzene dimer **PD** structure. In actual fact, all quantitative analyses show that the formation of stacking conformations can be sufficiently well explained by a combination of electrostatic and London dispersion interactions.

The examples chosen so far have always involved non-covalent interactions between just two binding partners. In solution chemistry this number is bound to increase simply due to the presence of solvent molecules or further equivalents of the interacting systems. How this impacts the non-covalent interactions presented above is best known for hydrogen-bonding interactions, and most specifically for water clusters of various sizes. In Figure 2.33 we see the structures of water dimer, trimer and tetramer optimized with an appropriate

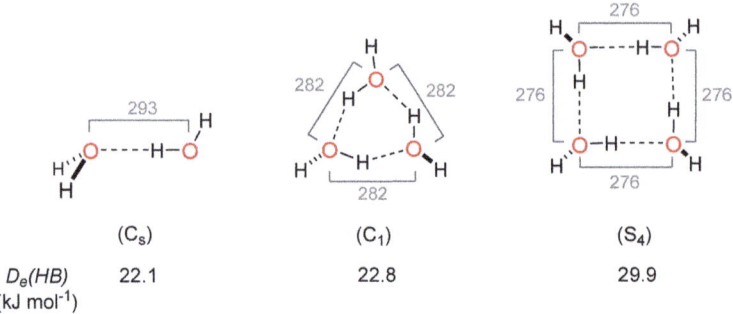

Figure 2.33 Structures (O···O distances, in pm) and dissociation energies
per hydrogen bond (D_e(HB), in kJ mol^{-1}) for the water dimer,
trimer and tetramer calculated at the CCSD(T)/aug-cc-pVDZ
level of theory.[29]

theoretical method. The first thing we note when comparing these
three systems is that the O–O bond distances get shorter (that is, the
cluster gets more compact) with increasing size. The energetic effort
required to dissociate the water clusters into individual water molecules will, of course, also increase with system size simply due to the
larger number of non-covalent interactions (in this case hydrogen
bonds) between the monomers involved. How the hydrogen bond
strengths in the three systems compare can thus be seen better by
dividing the overall binding energy by the (formal) number of hydrogen bonds in the cluster, to give the dissociation energy per hydrogen
bond (D_e(HB)). For the water dimer this energy amounts to D_e(HB)
$(H_2O)_2$ = 22.1 kJ mol^{-1}, which is more strongly binding than the BDE
value we have seen before for the same system (mainly because of the
different energy definitions employed). Increasing the cluster size we
note that the dissociation energy per hydrogen bond increases to a
value of D_e(HB)$(H_2O)_4$ = 29.9 kJ mol^{-1} for the water tetramer. Increasing the number of interacting water molecules thus shortens and
strengthens the intermolecular hydrogen bonds. This can most easily
be understood as a polarization effect of one water molecule onto its
direct neighbors, which in turn become better hydrogen bond donors
or acceptors in their interactions with other binding partners. This
phenomenon extends quite generally to bifunctional molecules capable of simultaneously acting as hydrogen bond donors and acceptors
and is often referred to as **polarization-enhanced hydrogen bonding**.
A complementary, and even more general, view of this observation is
that non-covalent interactions can quite generally enhance each other

in multi-component systems such that the overall binding energy of the system far exceeds the value expected from simple additive considerations. This is sometimes also referred to as a **cooperativity effect** among multiple non-covalent interactions.

In all of the above considerations we have neglected the influence of **entropic effects** on binding events. These become quite important in systems composed of a larger number of individual molecules, such as a single comparatively apolar organic solute surrounded by protic solvent molecules. In the extreme case, this may be a single benzene molecule surrounded by water molecules in a dilute aqueous solution. The first solvation shell of our benzene solute is more ordered than the bulk solvent and is composed of water molecules forming an insufficient number of hydrogen bonds to their direct neighbors. When two such benzene solutes react to form one of the dimer structures shown in Figure 2.32, a notable number of waters in the first solvation shell is liberated into bulk solution simply because the surface of the dimer is smaller than the combined surfaces of the monomers. The small binding energies we have seen for the benzene dimers before will not be sufficient to offset the entropic penalty for the loss of translational freedom. However, the water molecules liberated from the more ordered first solvation shell into bulk solvent provide a sufficient entropic as well as enthalpic driving force for the overall process to be favorable. This is sometimes referred to as the **hydrophobic effect** (or more generally, the **solvophobic effect**) and is considered to provide the main driving force for self-assembly processes of apolar solutes in polar solvents. It also provides a qualitative model for the limited solubility of organics in polar solvents.

2.8.2 Solvent Polarity Scales Using Molecular Probes

One of the most frequently employed general polarity scales for organic solvents has been developed by C. Reichardt and uses the betaine **A** shown in Figure 2.34a as a molecular probe. As is already apparent from the zwitterionic Lewis structure for **A**, this compound is characterized by a large dipole moment along the vertical axis and thus by a comparatively large solvation energy. The first excited singlet state of this compound can be described approximately as the result of a charge-neutralizing electron transfer from the formal phenolate unit to the central pyridinium ring, as expressed through Lewis structure **B**. This latter structure has a significantly smaller molecular dipole moment compared with **A** and thus also a much smaller solvation energy. These differences between the electronic ground and first

(a)

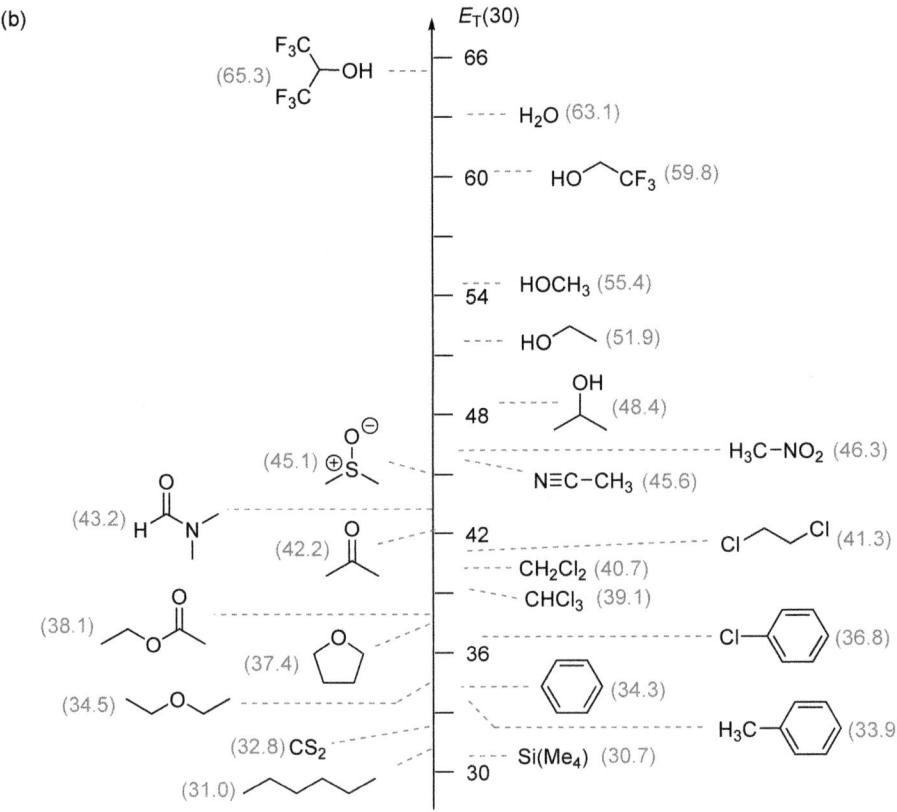

(b)

Figure 2.34 (a) Betaine dye **A** used as a molecular probe for determining the solvent polarity parameter $E_T(30)$. (b) $E_T(30)$ values for selected solvents.[30,31]

excited states **A/B** make the respective absorption energy strongly solvent dependent. In polar protic solvents, such as water, the excitation energy is comparatively large due to the much stronger stabilization of the ground state **A** compared to the excited state **B**. The respective solvent polarity parameter $E_T(30, \text{water}) = 63.1$ is then nothing else but the excitation energy measured in this solvent expressed in energy units of kcal mol^{-1}. In his initial work, Reichardt studied the solvent dependence of a larger series of betaine dyes, and the compound labeled "30" in this initial study is the one shown in Figure 2.34a and also the most generally applicable (which led to the "30" in the $E_T(30)$ polarity scale).

From the values shown in Figure 2.34b we can see that the most polar medium is hexafluoroisopropanol (HFIP) with $E_T(30, \text{HFIP}) = 65.3$, while the lower end of the scale is marked by tetramethylsilane (SiMe$_4$) and *n*-hexane with $E_T(30)$ values of 30.7 and 31.0, respectively. The most frequently employed organic solvents are those of intermediate polarity, as demonstrated, for example, by the $E_T(30)$ values of toluene (33.9), tetrahydrofuran (THF, 37.4), dichloromethane (DCM, 40.7) and acetone (42.2). Protic solvents capable of acting as hydrogen bond donors to the phenolate unit in betaine **A** are typically found in the upper half of the polarity scale shown here. $E_T(30)$ values of common solvents are sometimes also expressed in the form of their normalized (dimensionless) versions E_T^N calculated according to eqn (2.36). This involves taking the difference between the $E_T(30)$ value of a given solvent and the $E_T(30)$ value of tetramethylsilane ($E_T(30, \text{TMS}) = 30.7$), divided by the difference of the $E_T(30)$ values of TMS and water ($E_T(30, \text{water}) = 63.1$). This latter difference amounts to 32.4 and is taken as the polarity range accessible to common solvents with $E_T^N(\text{TMS}) = 0.00$ and $E_T^N(\text{water}) = 1.00$.

$$E_T^N(\text{solvent}) = \frac{E_T(\text{solvent}) - 30.7}{32.4} \tag{2.36}$$

where the normalized $E_T(30)$ value is $E_T^N(\text{solvent})$ and the $E_T(30)$ value of solvent is $E_T(\text{solvent})$.

Triethylphosphane oxide has been used by V. Gutmann as a molecular probe to measure the Lewis acidity of solvents. The Lewis structure shown for this compound in Figure 2.35a demonstrates that its formally negatively charged oxygen atom can readily interact with any Lewis acidic site in the surrounding solvent molecules. The fact that this may also include hydrogen-bonding interactions with the C–H bond in chloroform, is shown as an example in Figure 2.35a. The

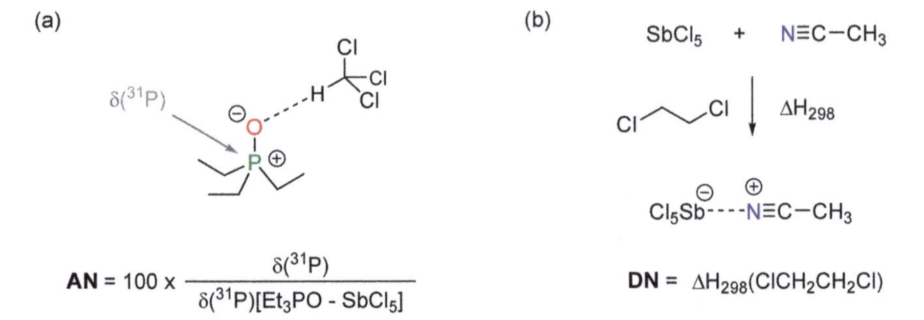

(a)

(b)

$$SbCl_5 \quad + \quad N\equiv C-CH_3$$

$\delta(^{31}P)$

$$Cl_5Sb\text{----}N\overset{\oplus}{\equiv}C-CH_3$$

$$\mathbf{AN} = 100 \times \frac{\delta(^{31}P)}{\delta(^{31}P)[Et_3PO - SbCl_5]}$$

$$\mathbf{DN} = \Delta H_{298}(ClCH_2CH_2Cl)$$

Figure 2.35 Definitions of (a) the Gutmann acceptor number (AN) for solvent Lewis acidity and (b) the Gutmann donor number (DN) for solvent Lewis basicity.[32]

Table 2.2 Gutman acceptor and donor numbers (AN and DN) for selected solvents.

Substituent	AN	DN
n-Hexane	0.0 (reference)	—
Diethylether	3.9	19.2
THF	8.0	20.0
Benzene	8.2	0.1
CCl_4	8.8	1.3
Ethyl acetate	9.3	14.8
Dioxane	10.8	15.0
Acetone	12.5	17.0
DMF	16.0	26.6
Acetonitrile	18.9	14.1
DMSO	19.3	29.8
CH_2Cl_2	20.4	—
$CHCl_3$	23.1	3.5
Isopropanol	33.5	21.1
Ethanol	37.1	18.5
Methanol	41.3	19.0
H_2O	54.8	18.0
$SbCl_5$	100 (reference)	—

NMR chemical shift of the phosphorus atom ($\delta(^{31}P)$) in triethylphosphane oxide provides a convenient analytical tool for quantifying the strength of this interaction. Comparing the measured chemical shifts for given solvents with those obtained for $SbCl_5$ as the (strong) reference Lewis acid gives, according to the equation shown in Figure 2.35a, the solvent "acceptor number" (AN) as a measure of the solvent Lewis acidity. As can be seen from the values collected in Table 2.2 for selected solvents, lower numbers correspond to lower solvent Lewis acidities and *n*-hexane is used to define the lower end of this scale with AN = 0.0. At the other end of the scale we find the reference Lewis acid $SbCl_5$ (AN = 100, by definition), protic solvents, such as water, methanol and ethanol, and then C–H acidic solvents, such as chloroform.

The Lewis acid $SbCl_5$ has also been employed by Gutmann to define the Lewis basicity of the solvent according to the concept illustrated in Figure 2.35b. As shown for acetonitrile as an example, the Lewis basicity of the solvent (also called the "donor number" (DN)) corresponds to the enthalpy of binding to the reference Lewis acid in dilute 1,2-dichloroethane solution (in kcal mol^{-1}). This property can, of course, only be measured in a reasonable way for solvents that have sufficiently Lewis-basic sites. From the values listed in Table 2.2 we see that DMF and DMSO are the most Lewis-basic solvents considered here, followed by aliphatic alcohols and aliphatic ethers (*e.g.* THF).

References

1. J. Helberg, M. Marin-Luna and H. Zipse, *Synthesis*, 2017, **49**, 3460.
2. A. G. Schultz and Y. K. Yee, *J. Org. Chem.*, 1976, **25**, 4045.
3. R. Loos, S. Kobayashi and H. Mayr, *J. Am. Chem. Soc.*, 2003, **125**, 14126.
4. M. Breugst, H. Zipse, J. P. Guthrie and H. Mayr, *Angew. Chem., Int. Ed.*, 2010, **49**, 5165.
5. R. G. Parr and R. G. Pearson, *J. Am. Chem. Soc.*, 1983, **105**, 7512.
6. R. G. Pearson, *Inorg. Chem.*, 1988, **27**, 734.
7. R. G. Pearson, *J. Org. Chem.*, 1989, **54**, 1424.
8. C. Hansch, A. Leo and R. W. Taft, *Chem. Rev.*, 1991, **91**, 165.
9. C. I. Ingold and W. S. Nathan, *J. Chem. Soc.*, 1936, 222.
10. H. C. Brown and Y. Okamoto, *J. Chem. Soc.*, 1958, **80**, 4979.
11. H. Jangra, Q. Chen, E. Fuks, I. Zenz, P. Mayer, A. R. Ofial, H. Zipse and H. Mayr, *J. Am. Chem. Soc.*, 2018, **140**, 16758.
12. H. Mayr, T. Bug, M. F. Gotta, N. Hering, B. Irrgang, B. Janker, B. Kempf, R. Loos, A. R. Ofial, G. Remennikov and H. Schimmel, *J. Am. Chem. Soc.*, 2001, **123**, 9500.
13. B. Kempf and H. Mayr, *Chem. – Eur. J.*, 2005, **11**, 917.
14. F. Brotzel, B. Kempf, T. Singer, H. Zipse and H. Mayr, *Chem. – Eur. J.*, 2007, **13**, 336.
15. H. Mayr, S. Lakhdar, B. Maji and A. R. Ofial, *Beilstein J. Org. Chem.*, 2012, **8**, 1458.
16. K. N. Houk, J. Sims, C. R. Watts and L. J. Luskus, *J. Am. Chem. Soc.*, 1973, **95**, 7301.
17. K. J. Morgan and D. P. Morrey, *Tetrahedron*, 1966, **22**, 57.
18. H. Fischer and L. Radom, *Angew. Chem., Int. Ed.*, 2001, **40**, 1340.
19. H. C. Brown, O. H. Wheeler and K. Ichikawa, *Tetrahedron*, 1957, **1**, 214.
20. M. Vasiliu, S. Li, K. A. Peterson, D. Feller, J. L. Gole and D. A. Dixon, *J. Phys. Chem. A*, 2010, **114**, 4272.
21. Y.-R. Luo, *Comprehensive Handbook of Chemical Bond Energies*, CRC Press, 2007.
22. J. A. Barker, R. A. Fisher and R. O. Watts, *Mol. Phys.*, 1971, **21**, 657.
23. A. Mukhopadhyay, W. T. S. Cole and R. J. Saykally, *Chem. Phys. Lett.*, 2015, **633**, 13.
24. A. Mukhopadhyay, S. S. Xantheas and R. J. Saykally, *Chem. Phys. Lett.*, 2018, **700**, 163.
25. M. D. Tissandier, K. A. Cowen, W. Y. Feng, E. Gundlach, M. H. Cohen, A. D. Earhart, J. V. Coe and T. R. Tuttle Jr., *J. Phys. Chem. A*, 1998, **102**, 7787.

26. J. C. Amicangelo and P. B. Armentrout, *J. Phys. Chem. A*, 2000, **104**, 11420.
27. M. O. Sinnokrot and C. D. Sherrill, *J. Phys. Chem. A*, 2006, **110**, 10656.
28. M. Pitonak, P. Neogrady, J. Rezac, P. Jurecka, M. Urban and P. Hobza, *J. Chem. Theory Comput.*, 2008, **4**, 1829.
29. E. Miliordos, E. Apra and S. S. Xantheas, *J. Chem. Phys.*, 2013, **139**, 114302.
30. C. Reichardt, *Chem. Rev.*, 1994, **94**, 2319.
31. V. G. Machado, R. I. Stock and C. Reichardt, *Chem. Rev.*, 2014, **114**, 10429.
32. V. Gutmann, *Coord. Chem. Rev.*, 1976, **18**, 225.

3 Pericyclic Reactions

Pericyclic reactions involve a reorganization of the bonding system in a closed cyclic arrangement of orbitals such that all bond-making and bond-breaking processes occur in a single kinetic step. All pericyclic reactions are therefore concerted, and proceed through a single transition state between reactants and products without formation of intermediates, such as biradicals, zwitterions or ion pairs. Pericyclic reactions are often subdivided into classes based on topological criteria and the numbers of σ- and π-bonds involved in the bond-making/breaking process. Typical subclasses are **electrocyclic reactions**, where acyclic reactants with n π-bonds convert to cyclic products with $(n-1)$ π-bonds and one new σ-bond. A typical example is the cyclization of hexa-1,3,5-triene to cyclohexa-1,3-diene shown in Figure 3.1a. The closely related class of **sigmatropic rearrangements** involve intramolecular migration of a σ-bond along a framework of π-bonds such that the overall number of σ- and π-bonds is retained in the process. A typical example is the "Cope rearrangement" of hexa-1,5-diene shown in Figure 3.1b. **Cycloaddition reactions** are ring-forming reactions involving the transformation of two reactants with a total of n π-bonds into cyclic products with $(n-2)$ π-bonds and two new σ-bonds. The Diels–Alder reaction of buta-1,3-diene with ethylene shown in Figure 3.1c captures the essence of this process. **Ene reactions**, as the last class involve the addition of one component with n π-bonds and one σ-bond to a second component with m π-bonds, such that a single product with $(n + m - 1)$ π-bonds and two σ-bonds are formed. The (formal) reaction of propene with ethylene to pent-1-ene shown in Figure 3.1d is a typical example.

Reactivity and Mechanism in Organic Chemistry, 2nd Edition
By Hendrik Zipse
© Hendrik Zipse 2023
Published by the Royal Society of Chemistry, www.rsc.org

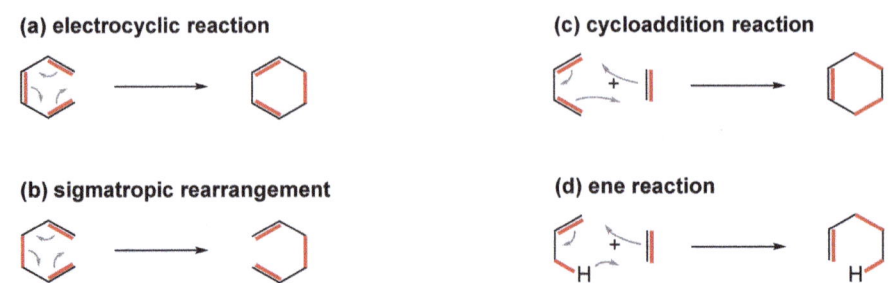

(a) electrocyclic reaction

(c) cycloaddition reaction

(b) sigmatropic rearrangement

(d) ene reaction

Figure 3.1 Prototypical examples for (a) electrocyclic reactions, (b) sigmatropic reactions, (c) cycloaddition reactions and (d) ene reactions.

While the arrow-pushing mnemonic shown in Figure 3.1 is helpful for tracking the fate of all bonds involved, a more detailed analysis of the molecular orbitals is required to understand the observed stereochemistry, as well as the large reactivity differences between seemingly related transformations. The basis for this type of analysis was laid by R. B. Woodward and R. Hoffmann, who discovered that the **conservation of orbital symmetry** is a key principle in all pericyclic reactions.[1] How this allows the characterization of pericyclic reactions as either being **symmetry allowed** or **symmetry forbidden** will be demonstrated in the following sections.

3.1   Electrocyclic Reactions

Electrocyclic reactions are most often categorized by the number of π-electrons actively involved in the reaction. The two largest classes are those involving four or six π-electrons (such as the example shown in Figure 3.1a). We will therefore use the class of 4π-electrocyclizations to follow the transformation of reactant to product orbitals and to see how orbital symmetry arguments apply.

3.1.1   Electrocyclic Reactions in Systems With Four π-Electrons

Thermal activation of *trans*-3,4-dimethylcyclobutene induces electrocyclic ring opening to (*E,E*)-hexa-2,4-diene (Figure 3.2). Comparison of reactant and product structures indicates that this transformation actively involves one π- and one σ-bond on the reactant side, which formally turn into two π-bonds (and four π-electrons) in the product. Photochemical activation of this product with 250 nm UV irradiation

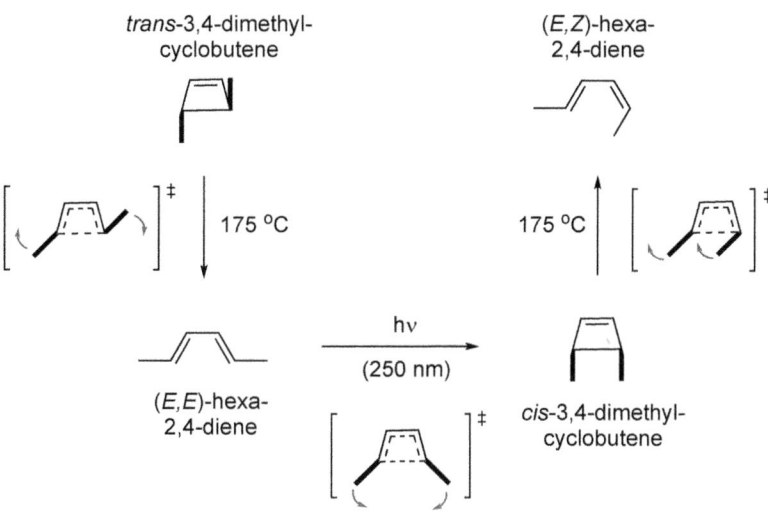

Figure 3.2 4π-Electrocyclizations/electrocyclic ring-opening reactions in the cyclobutene/buta-1,3-diene system.

leads to electrocyclic ring closure to *cis*-3,4-dimethylcyclobutene. Heating this product to 175 °C then triggers renewed electrocyclic ring opening, which now yields exclusively the (*E,Z*)-hexa-2,4-diene product. The results obtained in the thermal activation of the two stereochemically distinct cyclobutenes observed here imply a **stereospecific** process, whose reaction mechanism dictates the observed stereochemistry.

The stereochemical outcome of the ring-opening reaction of *trans*-3,4-dimethylcyclobutene shown in Figure 3.2 can be understood assuming a transition state in which the two methyl substituents move in a clockwise fashion away from the positions in the reactant towards the in-plane positions found in the diene product. That both termini involved in the reaction move in the same direction is termed **conrotatory** movement. This stereochemical observation also holds for the second ring-opening reaction shown in Figure 3.2, where conrotatory ring opening of *cis*-3,4-dimethylcyclobutene yields the (*E,Z*)-diene product. In contrast, the photochemically induced ring-closure reaction of the (*E,E*)-diene requires a transition state in which one end of the diene system moves clockwise and the other one counterclockwise in such a way that the two methyl substituents end up on the same side of the cyclobutene ring. This type of relative movement of the two termini is called **disrotatory**. A closer look at the orbital transformations accompanying the first of the above reactions, the electrocyclic ring-opening reaction of *trans*-3,4-dimethylcyclobutene, will show that only conrotatory ring opening is orbital symmetry allowed under thermal conditions for systems involving four π-electrons. For the sake of consistency with all other electrocyclic reactions we will analyze the reaction as a ring-closure reaction (from the product to the reactant side), as shown in Figure 3.3. In a first step we draw up the orbitals actively involved in the electrocyclic ring-closure reaction. On the acyclic side these are the four molecular orbitals MO1–MO4 of the diene π-system, and on the product side the two π-orbitals (π1 and π2) of the product C–C double bond, together with the σ/σ* orbitals of the newly formed single bond. In a second step we classify the reactant and product orbitals as symmetric (S) or antisymmetric (A) with respect to the C_2 axis, as the only symmetry element conserved along the conrotatory reaction pathway. What the **conservation of orbital symmetry** in pericyclic reactions now postulates is

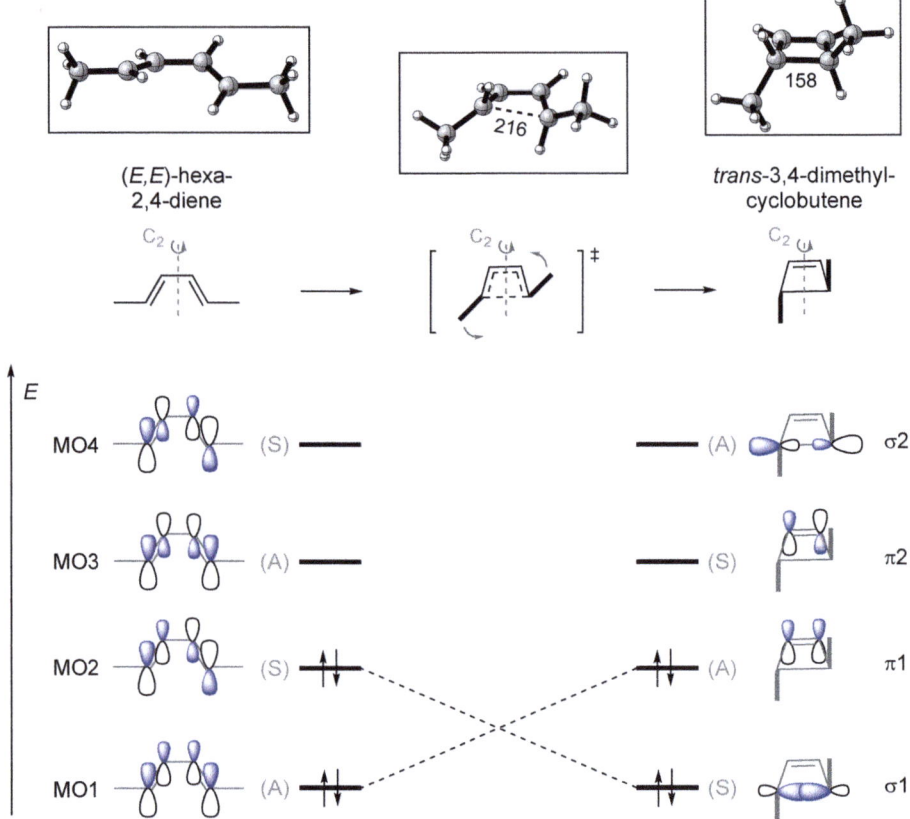

Figure 3.3 Orbital analysis for the conrotatory 4π-electrocyclization of (*E,E*)-hexa-2,4-diene, together with 3D structures of reactant, transition state and product. Selected distances are given in pm.

that electrons remain in an orbital of constant symmetry all along the reaction coordinate. The two electrons located in the lowest lying molecular orbital of A-type on the reactant side (MO1) will thus end up in the lowest lying molecular orbital of A-symmetry on the product side (π1). In an analogous manner it can be predicted that the two electrons located in S-type MO2 on the reactant side will end up in the lowest lying product orbital of S-symmetry (σ1). The electrocyclic reaction is called **symmetry allowed** if the procedure outlined above for connecting reactant and product orbitals of like symmetry generates the electronic ground state of the product. This is indeed the case for the conrotatory 4π-electrocyclization shown in Figure 3.3 (and for all other electrocyclizations involving 4π-electrons).

For the sake of comparison let us apply the same procedure to the disrotatory ring closure of the (*E,E*)-hexa-2,4-diene (Figure 3.4). The orbitals actively involved in this transformation are actually identical to those in the conrotatory electrocyclization discussed before. However, the classification of these orbitals according to their symmetry properties differs now, as the only symmetry element retained along the reaction pathway is the vertical mirror plane (σ_v)

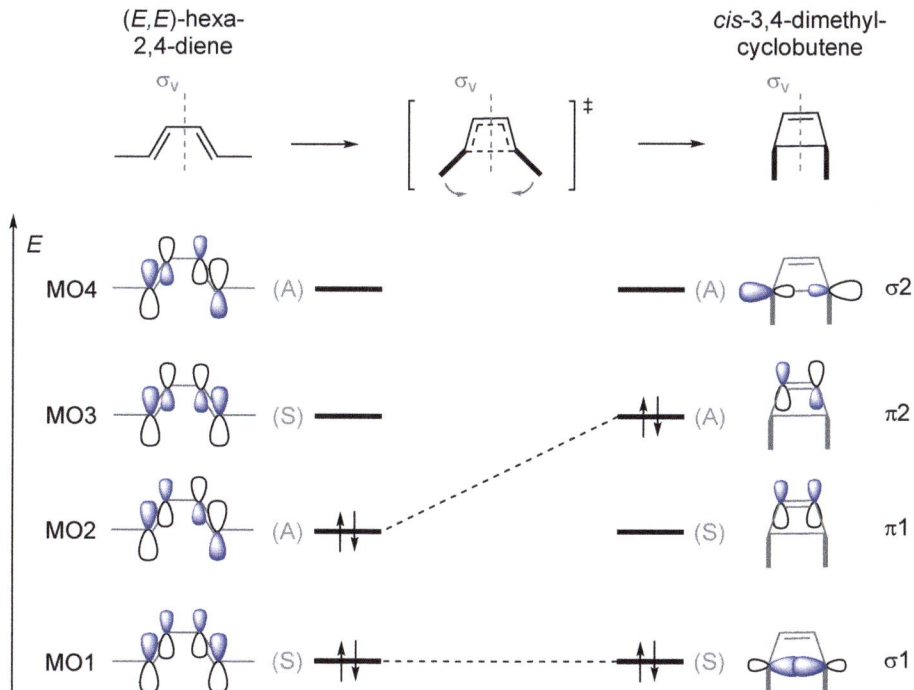

Figure 3.4 Orbital analysis of the disrotatory 4π-electrocyclization of (*E,E*)-hexa-2,4-diene.

cutting through the middle of the diene/cyclobutene system. On the reactant side the four molecular orbitals MO1–MO4 of the diene π-system (which are still the same as before) now all change their symmetry designation, the lowest molecular orbital MO1 now becomes S-type. On the product side the two lowest orbitals σ1 and π1 are S-type, while the upper two molecular orbitals are of A-symmetry. Again placing four electrons into the lowest two molecular orbitals, MO1 and MO2, on the reactant side, and following their fate along the reaction pathway while retaining the symmetry classification of the orbitals, we see that one electron pair ends up in σ1 and the second in π2. This does not, of course, correspond to the electronic ground state of the cyclobutene product, and this disrotatory 4π-electrocyclization is therefore termed **symmetry forbidden**.

This is true as long as we start from the electronic ground state of the reactant (which is, after all, what is meant if we assume "thermal conditions"). As shown in Figure 3.2 the disrotatory 4π-electrocyclization of (*E,E*)-hexa-2,4-diene proceeds readily upon photochemical activation. What this does to the analysis shown in Figure 3.4 is that the reactant is not in its electronic ground state. The energetically most facile electronic excitation often corresponds to the HOMO/LUMO promotion of a single electron, as shown in Figure 3.5. Following the

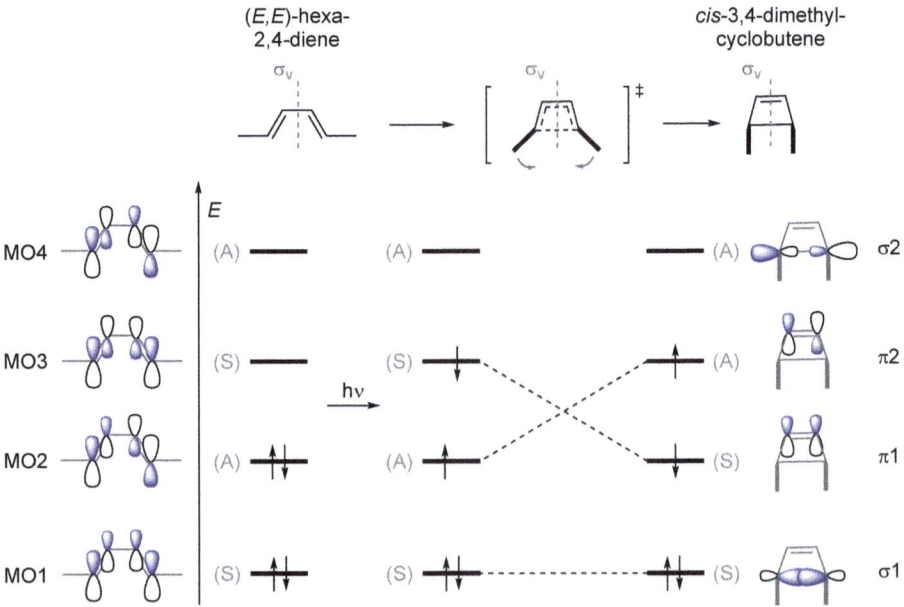

Figure 3.5 Orbital analysis of the photochemically induced disrotatory 4π-electrocyclization of (*E,E*)-hexa-2,4-diene.

fate of this electronic configuration from the reactant to the product side, and keeping all electrons again in orbitals of constant symmetry, we can see that the product electronic state is excited as much as was the case for the reactant side. Under this condition, the reaction is **symmetry allowed**. What complicates the analysis of photochemically induced reactions is the variety of electronic states that can be generated experimentally through UV–visible irradiation. The assumption of a HOMO/LUMO excitation, as shown in Figure 3.5, is no more than that, an assumption. On a more general note the orbital analyses shown in Figures 3.3–3.5 are simplistic in that they neglect orbital interactions between the core reaction system and the rest of the molecule (such as the methyl substituents).

The stereochemical selection rules described above have enormous consequences for the chemistry of annelated cyclobutenes. Many of these are much more stable than might be anticipated based on their ring strain energy. This is best seen in the properties of Dewar benzene derivatives, which can be synthesized photochemically from substituted benzenes. The cyclization of 1,2,4-tri-*tert*-butylbenzene (**A**), for example, leads to the respective Dewar-isomer 1,2,5-tri-*tert*-butylbicyclo[2.2.0]hexa-2,5-diene (**B**) (Figure 3.6). As indicated by the bonds marked in blue, this can be seen as a photochemically induced symmetry-allowed disrotatory 4π-electrocyclization. Dewar benzene **B** is highly strained, but stable at room temperature and requires thermal activation for ring opening back to **A**. This latter process is symmetry forbidden, under thermal conditions, and may even proceed along a stepwise pathway involving biradical intermediates. The symmetry-allowed conrotatory ring opening of **B** does not happen due to the high ring strain induced through the formation of a *trans* double bond in the (hypothetical) product **C**. Dewar benzene isomer **B** is thus trapped between the thermochemical disaster waiting in **C** and the symmetry-forbidden (high barrier) ring opening back to **A**.

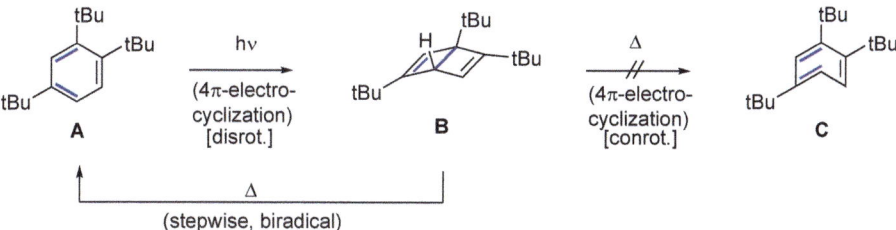

Figure 3.6 Photochemical cyclization of 1,2,4-tri-*tert*-butylbenzene (**A**) to Dewar benzene **B**.

In thermally allowed conrotatory processes substituents present at the 3- and 4-positions of cyclobutenes can either rotate inwards (forming alkenes with *Z* configuration) or outwards (to *E*-alkenes). While the formation of *E*-alkenes is generally preferred on thermochemical grounds, the formation of *Z*-alkenes is nevertheless observed for cyclobutenes carrying acceptor substituents. A well-studied example for this **torquoselectivity** is 3-formylcyclobutene, whose thermally induced electrocyclic ring-opening reaction yields (*Z*)-penta-2,4-dienal exclusively (Figure 3.7). This result is rationalized assuming a kinetic preference for inward rotation of the formyl group in the conrotatory transition state. On thermochemical grounds the respective (*E*)-isomer is more stable, a fact readily demonstrated through acid-catalyzed isomerization of the (*Z*)-penta-2,4-dienal reaction product. The inward rotational preference observed for electron-withdrawing substituents like the formyl group has been rationalized by K. N. Houk through favorable interactions between vacant acceptor orbitals (represented here by an empty p-type AO) and the HOMO of the reacting cyclobutene system. In the transition state the latter corresponds mainly to the σ(C–C) orbital of the breaking C–C bond between atoms C3 and C4, while the corresponding σ*(C–C) orbital corresponds to the overall LUMO. Donor–acceptor interactions between the C–C bond HOMO and the empty substituent orbital are more favorable in the inward as opposed to the outward transition state, simply due to a more favorable relative orientation (and thus better overlap).

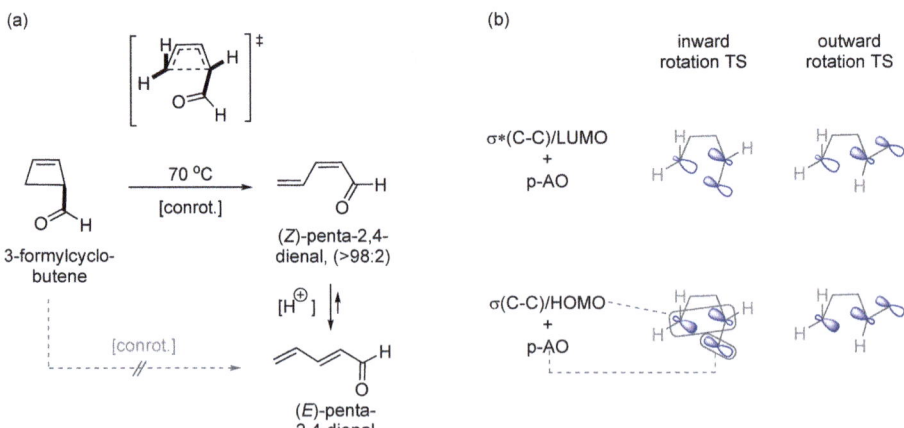

Figure 3.7 (a) Torquoselectivity in the ring-opening reaction of 3-formylcyclobutene, and (b) frontier molecular orbital (FMO) rationalization of the rotational selectivity.[2]

The synthetically most relevant 4π-electrocyclization is the **Naza-rov cyclization** (Figure 3.8). This reaction involves initial activation of divinylketone substrates with a proton source (or Lewis acid) to yield a 3-hydroxypentadienyl cation. Conrotatory ring closure of this intermediate generates a five-membered ring structure containing an allyl cation fragment, whose deprotonation and subsequent keto/enol tautomerization then yields the final cyclopentenone product. The deprotonation step proceeds typically such that the more highly sub-stituted double bond is formed. In the example shown in Figure 3.8a this implies the loss of one of the stereocenters generated in the elec-trocyclization step. Analysis of the orbital symmetry requirements is most easily performed here on the parent pentadienyl cation system shown in Figure 3.8b. The five molecular orbitals required to describe the π-system of this species have already been discussed in Chapter 1, Section 1.3.1. On the product side of the reaction we combine the MOs of the allyl cation (also presented in Section 1.3.1) with σ/σ* orbitals of the newly formed C–C single bond. All reactant and prod-uct orbitals are then classified according to the C_2 rotational axis as the principal symmetry element retained along the conrotatory reac-tion coordinate. Filling the four π-electrons into the lowest reactant MOs and following their fate along the reaction pathway, we see that

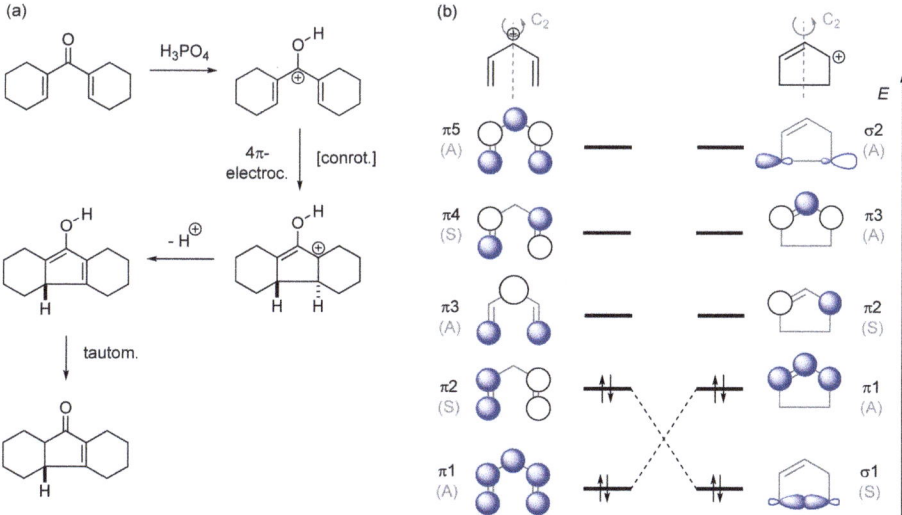

Figure 3.8 (a) The acid-catalyzed Nazarov cyclization of divinylketones and (b) orbital correlation diagram for the conrotatory 5-center/4π-electrocyclization of the pentadienyl cation.

the product is formed in its electronic ground state. The conrotatory cationic 5-center/4π-electrocyclization is therefore symmetry allowed.

The synthetic value of the Nazarov cyclization reaction can be extended by steering the direction of the deprotonation step (and thus the location of the C–C double bond). This can be achieved with the **silicon-directed Nazarov cyclization** (Figure 3.9), for example. The reaction utilizes β-silylvinylketones as substrates and works best with chloride-based Lewis acids. Activation and cyclization of the divinyl-ketone substrate proceeds as before to furnish a β-silyl-substituted allyl cation. Elimination of the silyl group, most likely aided by a Lewis acid-derived chloride ion, is now faster than the competing deprotonation steps and thus yields the enolate product with the less substituted double bond. After aqueous work-up the *cis*-annelated cyclopentenone is obtained in 84% yield as the only regio- and stereo-isomer. The pentadienyl cation intermediates generated in the Naza-rov cyclization reaction are also accessible from other precursors, the most relevant being the respective pentadienyl halides or alcohols. An example of the latter case can be found in the synthesis of 1,2,3,4,5-pentamethylcyclopentadiene, whose deprotonation yields the pen-tamethylcyclopentadienyl anion (Cp*) ligand frequently employed in transition metal chemistry (Figure 3.10). Initial reaction of ethyl acetate with two equivalents of (*Z*)-but-2-en-2-yllithium yields, after aqueous work-up, the pentadienyl alcohol skeleton carrying all five methyl groups already in the required positions. Acid-catalyzed elimi-nation of the alcohol OH group then generates the pentadienyl cation intermediate, whose 5-center/4π-electrocyclization and subsequent deprotonation yields the final 1,2,3,4,5-pentamethylcyclopentadiene product.

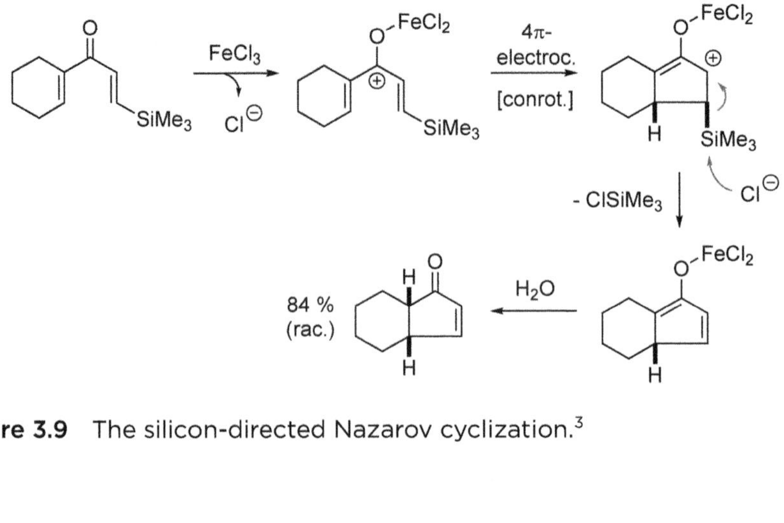

Figure 3.9 The silicon-directed Nazarov cyclization.[3]

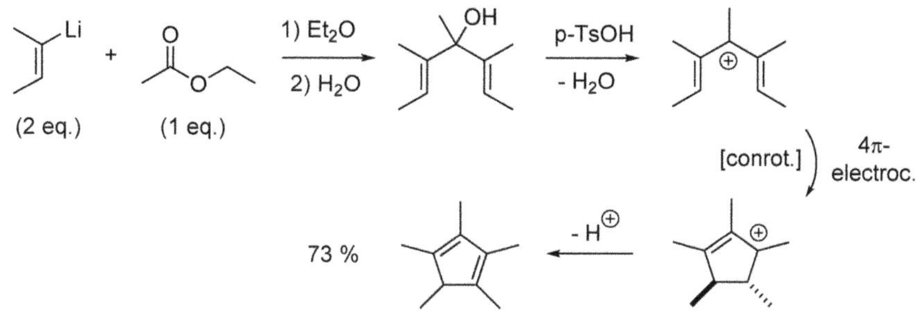

Figure 3.10 Synthesis of 1,2,3,4,5-pentamethylcyclopentadiene.[4]

3.1.2 Electrocyclic Reactions in Systems With Six π-Electrons

6π-Electrocyclization reactions show opposite rotational preferences compared to those involving four π-electrons. This can readily be seen from the reactions compiled in Figure 3.11. Thermal activation of (*E,Z,E*)-octa-2,4,6-triene thus leads to *cis*-5,6-dimethylcyclohexa-1,3-diene as the only product, indicative of a disrotatory ring-closure process. In contrast to the examples discussed for the 4-center/4π-electrocyclizations in Figure 3.2, the equilibrium between ring-opened and ring-closed forms is shifted to the side of ring closure in most 6π-electrocyclizations due to much lower ring strain energies. Photochemical activation of the same substrate as used before leads to ring closure with opposite (conrotatory) rotational preferences and thus to *trans*-5,6-dimethylcyclohexa-1,3-diene. This product is also accessible from (*Z,Z,E*)-octa-2,4,6-triene through disrotatory 6π-electrocyclization under thermal conditions. The stereochemical preferences observed under thermal conditions here can easily be rationalized using the same type of orbital correlation diagram as in 4π-electrocyclization reactions (Figure 3.12). The six π-MOs making up the triene π-system on the reactant side are paired up with the four π-MOs and the σ/σ^* orbitals on the product side. Classification of all orbitals is now with respect to the vertical mirror plane (σ_v) cutting through the middle of the triene/cyclohexadiene system. Following the six electrons located in MO1–MO3 from the reactant to the product side we arrive at the electronic ground state configuration of the cyclohexadiene system, and the disrotatory electrocyclization analyzed here is thus orbital symmetry allowed. The corresponding conrotatory alternative is symmetry forbidden, as can be easily shown by switching to the C_2 rotational axis as the symmetry element used

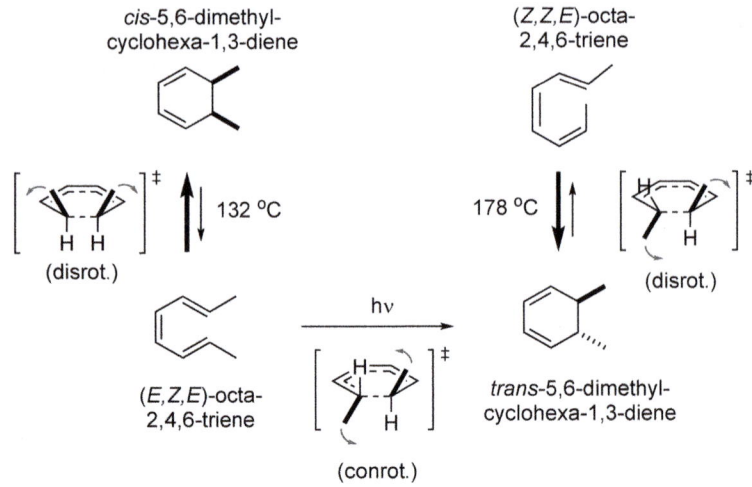

Figure 3.11 6π-Electrocyclizations/electrocyclic ring-opening reactions in the cyclohexadiene/hexa-1,3,5-diene system.[5]

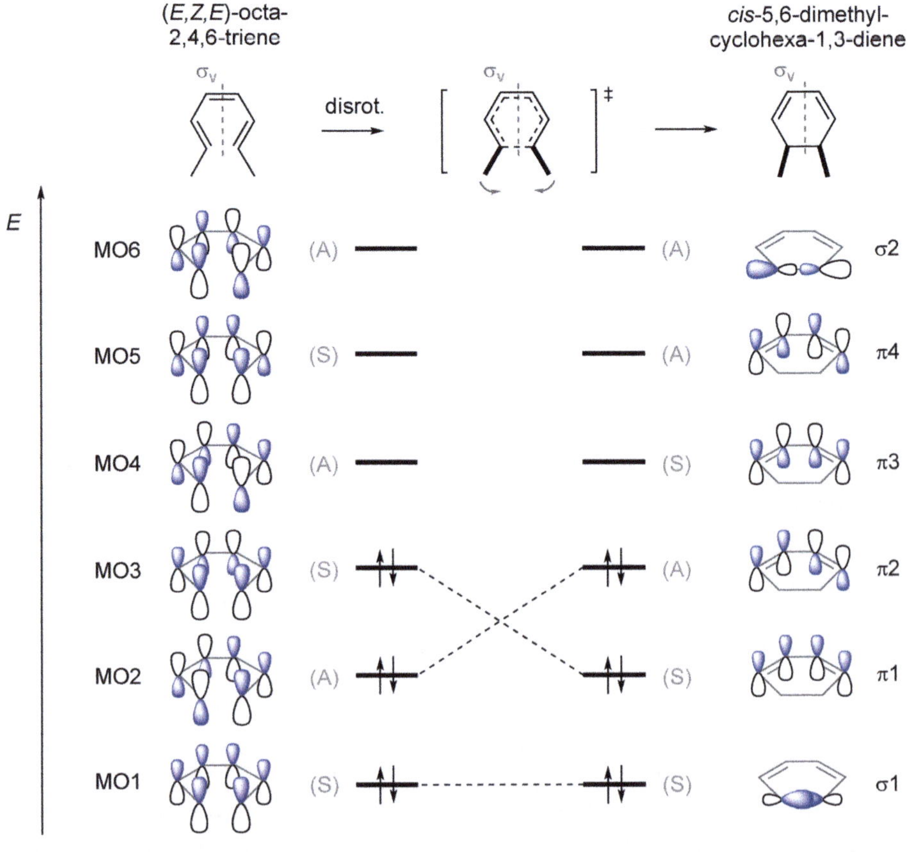

Figure 3.12 Orbital analysis of the disrotatory 6π-electrocyclization of (E,Z,E)-octa-2,4,6-triene.

for classification. Following the logic outlined in the analysis of photochemically promoted 4π-electrocyclizations in Section 3.1.1, it is also easy to show that conrotatory 6π-electrocyclizations are allowed under photochemical conditions. The orbitals required for this analysis are again those shown in Figure 3.12, and the guiding symmetry element is again the C_2 axis running through the center of the hexatriene/cyclohexadiene systems.

The photochemically induced conrotatory 6π-electrocyclization/ring opening plays a major role in the biosynthesis of vitamin D_3 (Figure 3.13). The actual substrate for this photochemical reaction, 7,8-dehydrocholesterol, is initially produced through enzyme-mediated oxidation of cholesterol. The photochemically induced ring-opening reaction then occurs in the skin upon exposure to sunlight (UV-B irradiation at around 300 nm). The initially formed ring-opened substrate pre-vitamin D_3 then undergoes spontaneous isomerization through 1,7-sigmatropic hydrogen migration, to give vitamin D_3. The stereochemical requirements for this last step (antarafacial) will be discussed in Section 3.3.3. Vitamin D_3 is important in regulating calcium metabolism, and insufficient exposure to (sun)light therefore leads to insufficient bone mineralization in children (rickets).

Important variations of 6π-electrocyclizations include the 5-center/6π-electrocyclization of carbanions and their heteroatom analogs. A well-known example for carbanions is the base-induced cyclization of cycloocta-1,4-diene (Figure 3.14a). Reaction of this substrate with *n*-butyllithium in THF at −78 °C yields the corresponding cyclooctadienyl anion as a stable species amenable to

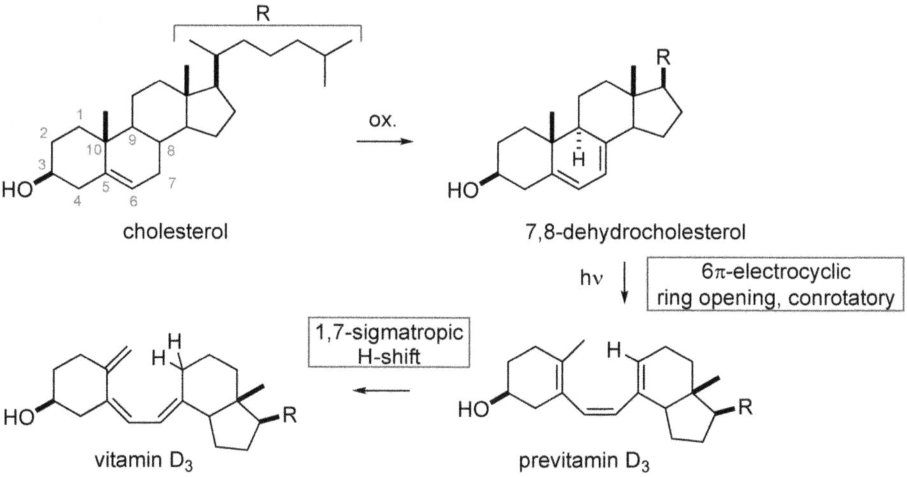

Figure 3.13 Biosynthesis of vitamin D_3 from cholesterol.

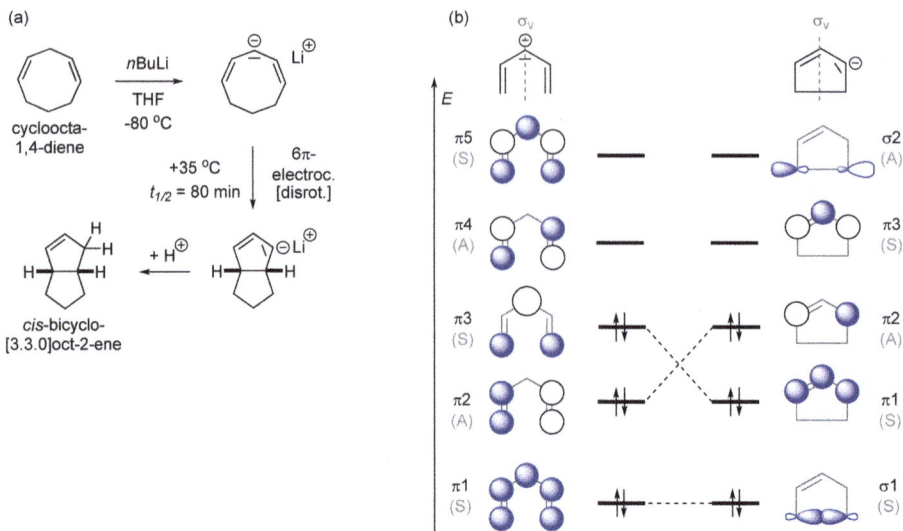

Figure 3.14 (a) 6π-Electrocyclization in cyclooctadienyllithium and (b) orbital correlation diagram for the disrotatory 5-center/6π-electrocyclization of the pentadienyl anion.[6]

NMR spectroscopic characterization. On warming the reaction mixture to +35 °C, 6π-electrocyclization occurs with a half-life of $t_{1/2}$ = 80 min. The final product *cis*-bicyclo[3.3.0]oct-2-ene is obtained after quenching with a proton source. In line with expectation for a thermal 5-center/6π-electrocyclization, the reaction proceeds in a disrotatory fashion. The orbital interaction diagram for the prototype of this reaction, the cyclization of the pentadienyl anion, is shown in Figure 3.14b. The orbitals required on the reactant and product side are actually identical to those appearing in the pentadienyl cation cyclization in Figure 3.8. However, the symmetry element retained along the disrotatory reaction pathway is the vertical mirror plane cutting through the pentadienyl anion ring system. In addition, the anionic case studied here involves six π-electrons on the reactant side of the diagram. Following these electrons to the product side generates the product electronic ground state and the reaction is thus orbital symmetry allowed.

Further variations of 6π-electrocyclizations derive from the pentadienyl anion cyclization through formal replacement of the carbanion center by neutral heteroatoms carrying lone-pair electrons. Easily accessible substrates of this type include hydrazones of α,β-unsaturated ketones, whose acid-induced cyclization leads to pyrazolines. That the use of chiral (phosphoric) acids leads to

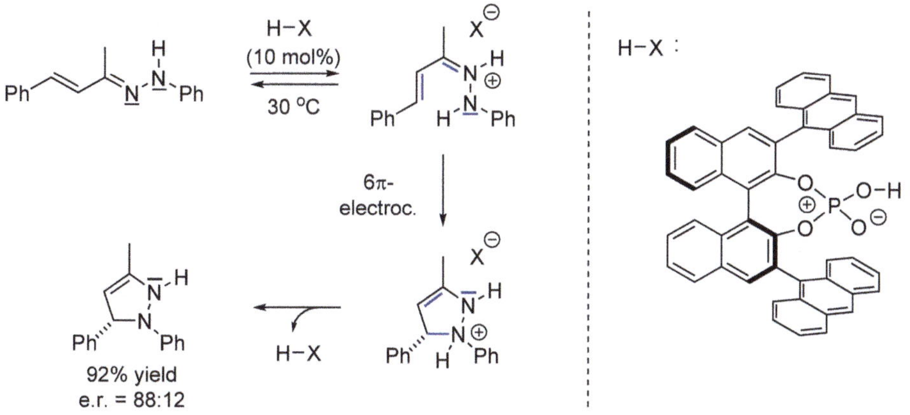

Figure 3.15 Stereoselective synthesis of 3-methyl-1,5-diphenyl-3-pyrazoline.[7,8]

stereochemical control of the cyclization reaction has recently been shown for substrates carrying aryl substituents (Figure 3.15). The chiral binolphosphoric acid catalyst employed here is assumed to protonate the substrate imine nitrogen atom and thus facilitate conformational reorientation such that the reactive C and N termini are positioned adjacent to each other. The subsequent bond-making/bond-breaking processes involve the formal C–C and C–N double bonds, together with the terminal nitrogen lone pair. Cyclization then occurs such that the (*S*) configuration product is formed preferentially (enantiomeric ratio, e.r. = 88:12) in 92% yield. The fact that the cyclization reaction proceeds in one single step has been confirmed by quantum chemical studies, together with the very substantial acceleration through acid catalysis. However, in contrast to carbocyclic analogs, as well as the (high barrier) cyclization of the unprotonated system, the reaction does not appear to fulfill all the criteria for pericyclic reactions, and may instead be better understood as a nucleophilic addition of the amine terminus to the α,β-unsaturated iminium fragment.

3.1.3 Electrocyclic Reactions in Systems With Two and Eight π-Electrons

Reviewing the examples for 4π- and 6π-electrocyclizations in the previous sections suggests that the number of π-electrons (before cyclization) and the preference for conrotatory or disrotatory ring closure are tightly interconnected. Additional consideration of systems with two or eight π-electrons leads to the systematic overview shown in Figure 3.16. The smallest systems are those involving allyl cations (at the ring-opened stage) and show a disrotatory rotational preference. On the other end of the scale we find the 8π-electrocyclization of octatetraenes with a conrotatory preference. While the rotational preference of all these systems can be analyzed following the orbital correlation diagram approach demonstrated before, analysis of the HOMO of the π-system at the ring-opened stage is usually sufficient: the rotational preference of the ring-closing electrocyclization is derived by turning the p-type atomic orbitals at the terminal positions such that like phases rotate towards each other. Counting electrons, we can also see that all systems with $(4n + 2)$ "active" electrons undergo disrotatory movement under thermal (that is, electronic ground state) conditions, while systems with $4n$ electrons prefer the conrotatory mode.

With these general rules in mind it is straightforward to predict the stereochemical course of the 8π-electrocyclization of

system	no. π-electrons	π-HOMO	rotational preference
	2		disrotatory
	4		conrotatory
	4		conrotatory
	4		conrotatory
	6		disrotatory
	6		disrotatory
	8		conrotatory

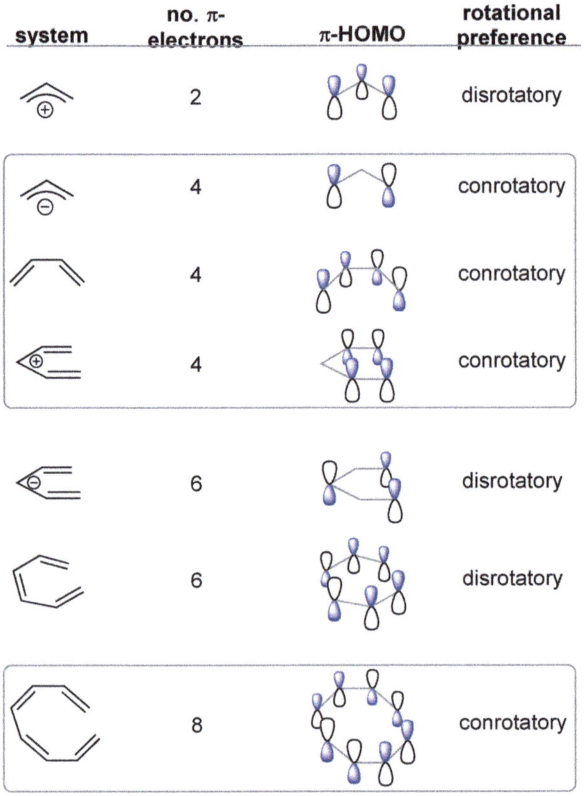

Figure 3.16 Electrocyclization reactions in different π-systems.

(2*E*,4*Z*,6*Z*,8*E*)-deca-2,4,6,8-tetraene shown in Figure 3.17a. This reaction proceeds at −10 °C with conrotatory movement of the reacting termini to *trans*-7,8-dimethylcycloocta-1,3,5-triene as the only reaction product. On warming to 20 °C this latter compound undergoes a second cyclization, this time of the 6π-electrocyclization type with disrotatory stereochemical preference. The fact that both cyclization reactions are stereospecific can be demonstrated by starting the reaction from the (2*Z*,4*Z*,6*Z*,8*E*)-deca-2,4,6,8-tetraene stereoisomer (Figure 3.17b). As must be expected from a conrotatory ring-closure process, the only reaction product formed is *cis*-7,8-dimethylcycloocta-1,3,5-triene. Heating to +40 °C triggers a second cyclization to *cis*-7,8-dimethylbicycloc[4.2.0]octa-2,4-diene through the same disrotatory 6π-electrocyclization observed for the other stereoisomer before.

Ring-opening reactions of cyclopropyl to allyl cations have been studied in detail as the prototype for 2π-electrocyclic ring-opening reactions. Experimental studies typically involve cyclopropyl halides or cyclopropanol esters, such as the tosylates (Tos) shown in Figure 3.18. Thermal activation of these highly strained precursors in solvolysis experiments leads to a combined ionization/ring-opening reaction, and yields the allyl cations in all cases with the expected disrotatory stereochemistry. The final products result from trapping the allyl cations with a nucleophile such as acetate. For the three substrates shown in Figure 3.18 the rate of reaction depends dramatically on the relative stereochemistry of ring substituents and the tosylate leaving group: *cis*,*cis*-2,3-dimethylcyclopropyl tosylate reacts slowest and generates the sterically biased *cis*,*cis*-dimethylallyl cation as the first intermediate (Figure 3.18a). The sterically less biased *cis*,*trans*-dimethylallyl cation is generated from the *cis*,*trans*-2,3-dimethylcyclopropyl tosylate

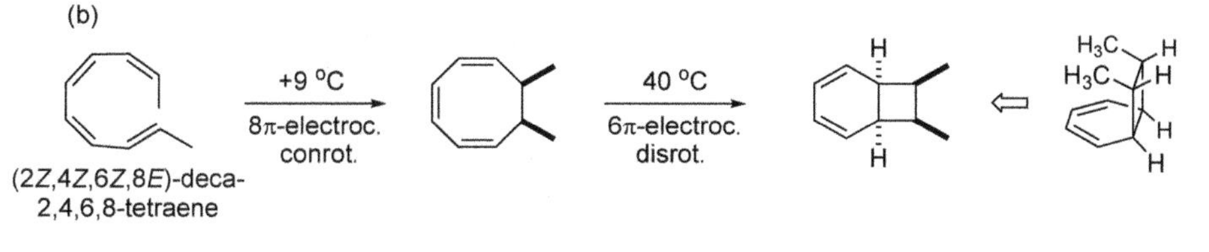

Figure 3.17 Electrocyclization cascades in deca-2,4,6,8-tetraene substrates.[9]

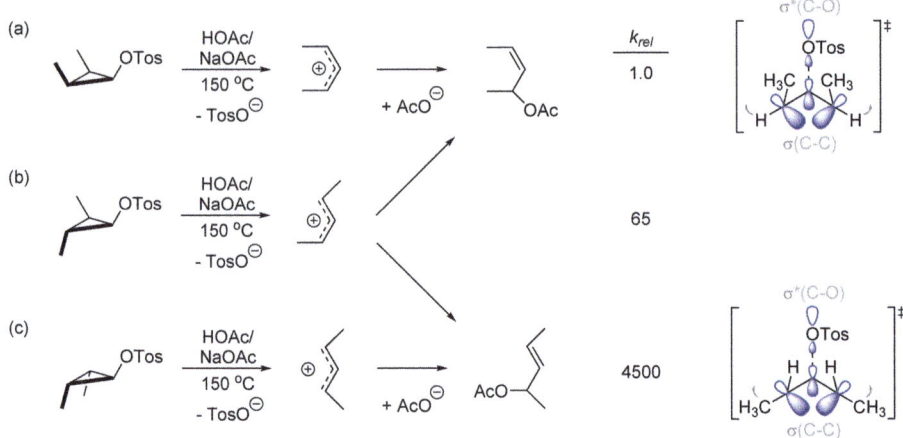

Figure 3.18 Solvolysis/electrocyclic ring opening of substituted cyclopropyl tosylates.[10,11]

under the same reaction conditions 65 times faster, as shown in Figure 3.18b. The fastest reaction rate (k_{rel} = 4500) is observed for *trans,trans*-2,3-dimethylcyclopropyl tosylate, whose ionization/ring-opening reaction yields the *trans,trans*-dimethylallyl cation exclusively. The torquoselectivity and the relative rates observed in the disrotatory ring-opening reactions in Figure 3.18 can be rationalized assuming strong donor/acceptor interactions between the σ(C–C) orbital of the breaking C–C bond and the σ*(C–O) orbital of the breaking C–OTos bond. In the reaction of the *cis,cis*-dimethylcyclopropyl tosylate, orbital overlap is best when the methyl substituents rotate inwards, as shown schematically in the transition state in Figure 3.18a. While this helps the actual ionization process, the transition state is burdened by steric repulsion between the methyl groups rotating inwards. This is not the case in the reaction of the *trans,trans*-dimethylcyclopropyl tosylate shown in Figure 3.18c, where the same favorable donor/acceptor interactions can be achieved by rotating the methyl substituents outwards (that is, away from each other).

3.1.4 Electrocyclic Reactions of Radicals

The survey of electrocyclic reactions in Figure 3.16 includes only closed-shell systems and the question naturally arises, whether the Woodward–Hoffmann rules also apply to open-shell species.

A well-known radical undergoing electrocyclic ring opening is the cyclopropyl radical **A** (Figure 3.19), whose ring opening to allyl radical

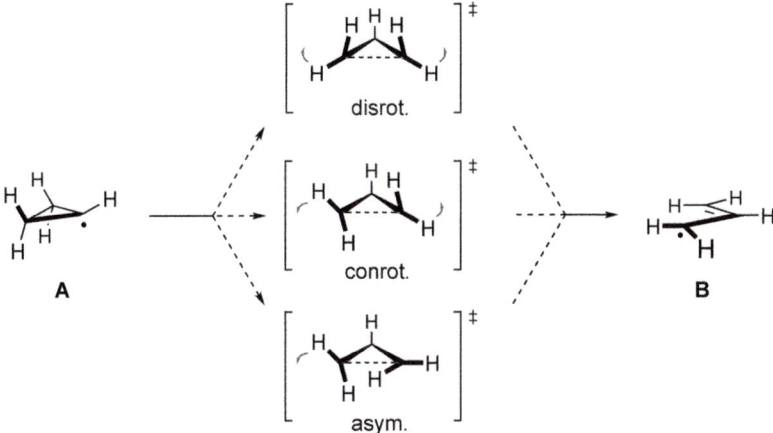

Figure 3.19 Ring-opening reaction of cyclopropyl radical **A** to allyl radical **B**.[12-14]

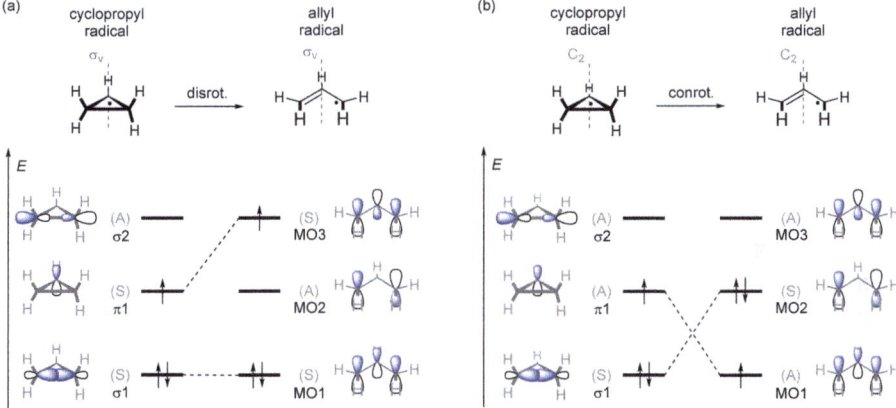

Figure 3.20 Disrotatory and conrotatory ring opening of cyclopropyl radical to allyl radical.[12-14]

B faces a barrier of only 92 kJ mol^{-1} and is exothermic by 95 kJ mol^{-1}. Whether this reaction proceeds in a conrotatory or disrotatory fashion is, unfortunately, not known experimentally. Following the same procedure as for the closed-shell system we can analyze the disrotatory process with the orbital correlation diagram shown in Figure 3.20a. The symmetry element retained along the disrotatory reaction pathway is the vertical σ_v mirror plane bisecting the breaking C–C bond and the radical center. Labeling all reactant and product orbitals according to this symmetry element, we find the three "active" electrons in the lower two orbitals, both of which are of (S) symmetry.

Retaining these symmetry designations from the reactant to the product side, we find that the allyl radical is generated in an excited electronic state. Reaction along a disrotatory pathway is thus forbidden according to the Woodward–Hoffmann rules.

The orbital correlation diagram for the alternative conrotatory ring-opening reaction is shown in Figure 3.20b. Conrotatory reaction pathways usually retain a twofold rotational axis along the reaction pathway. This meets with some problems here due to the fact that the radical center of the cyclopropyl radical is pyramidalized. The C_2 rotational axis is thus a valid symmetry element only for the (slightly artificial) assumption that the radical center is planar. However, going through the orbital correlation analysis with this assumption in mind, we find that the reaction is symmetry forbidden. In contrast to the closed-shell systems analyzed before, we therefore find that the disrotatory and conrotatory reaction pathways are both forbidden for the cyclopropyl radical—at least as long as they retain the same symmetry properties from reactants to products. Extensive quantum chemical studies using a variety of theoretical methods all agree that the cyclopropyl radical escapes the slavery of symmetry control and ring opens through a lower (that is, C_1) symmetric reaction pathway. The transition state can best be visualized as the one shown in Figure 3.19, where one of the allyl radical termini has rotated into its final position, while the other end is just at the beginning of this process.

3.1.5   Related Cyclization Reactions

Several cyclization reactions of extended π-systems exist with strong resemblance to one of the electrocyclic reactions discussed in the previous sections. One of these is the Bergman cyclization of enediynes. Heating deuterated enediyne **A**, Jones and Bergman observed rapid scrambling of deuterium between the 1,6- and 3,4-positions (Figure 3.21). This can be rationalized by assuming the symmetric *p*-benzyne intermediate **B**, whose reactivity can best be understood as that of a 1,4-biradical. This assumption is also supported by trapping reactions performed in the presence of toluene or CCl_4. While 1,4-dichlorobenzene-2,3-d_2 (**D**) is formed in the latter case, the reaction with toluene leads to selectively deuterated diphenylmethane **E** as the exclusive product. Best thermochemical estimates position *p*-benzyne **B** only 44 kJ mol^{-1} above enediyne **A**.

In substrates carrying radical-stabilizing substituents at the alkyne termini, the Bergman cyclization competes with the "Schreiner–Pascal" (that is, the 1,5-cyclization) reaction. This is exemplified by the

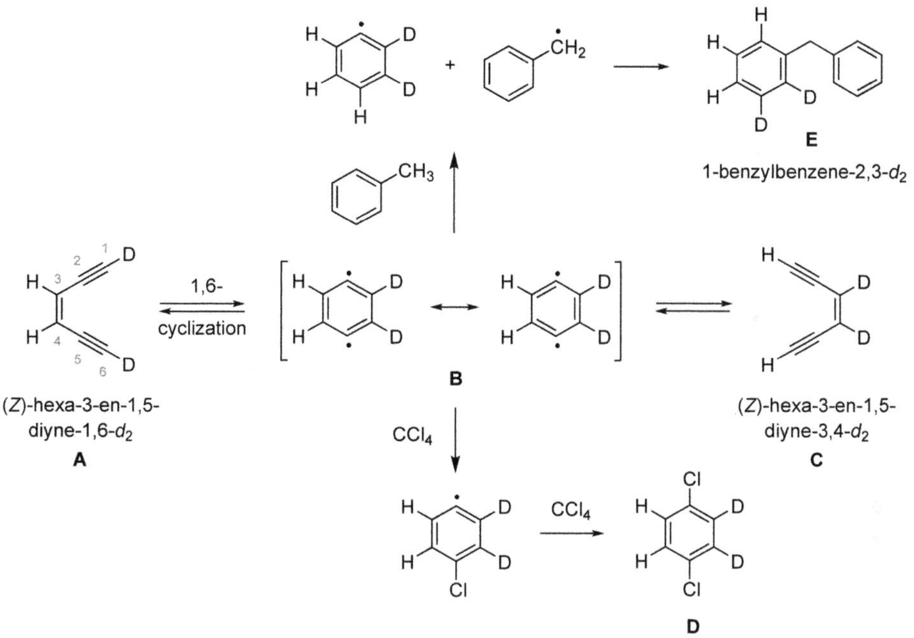

Figure 3.21 Bergman cyclization of (*Z*)-hexa-3-en-1,5-diyne-1,6-d$_2$ (**A**).[15,16]

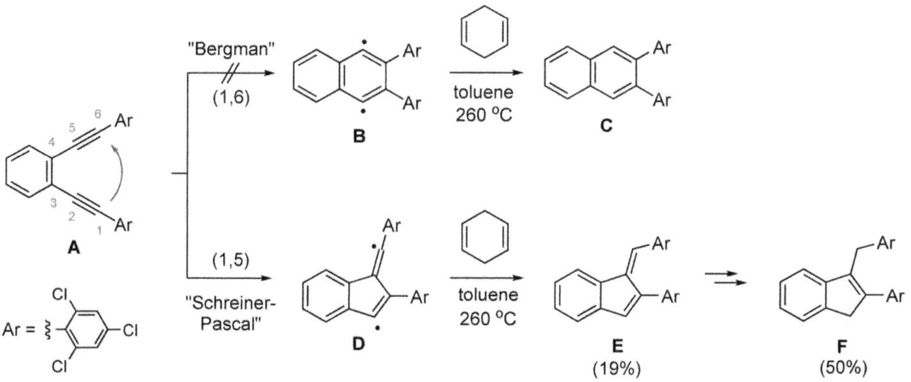

Figure 3.22 Competing Bergman and Schreiner–Pascal cyclizations.[17,18]

1,6-disubstituted substrate **A** in Figure 3.22, where the central alkene unit has been replaced by a benzene fragment. Heating this substrate to 260 °C in the presence of cyclohexa-1,4-diene as an H-atom donor yields none of the 2,3-disubstituted naphthalene product **C** expected after the Bergman cyclization, and the subsequent trapping by hydrogen atom transfer. Instead, the reaction through 1,5-cyclization to biradical **D** appears to be much faster due to stabilization of one of the unpaired spins through the attached aryl group. The product obtained after double hydrogen atom transfer (indene **E**, 19%) is, however, not stable under the reaction conditions and undergoes further reduction to the bicyclic product **F** (50%).

Closely related to the Bergman cyclization is the Myers–Saito cyclization of eneyne allenes (Figure 3.23). This reaction formally derives from the Bergman cyclization through replacement of one of the substrate triple bonds by an allene fragment. The prototypical hepta-1,2,4-triene-6-yne substrate (**A**) undergoes thermally induced cyclization to biradical **B**. In analogy to the Bergman cyclization, this cyclization mode is referred to as "1,6-cyclization", even though this may not be in line with standard IUPAC numbering conventions of the actual substrate. All trapping reactions of biradical **B** indicate that the two formally unpaired spins in **B** have dramatically different reactivities. Taking the reaction with cyclohexa-1,4-diene as a highly reactive hydrogen atom donor as an example, initial hydrogen transfer occurs to the radical center located directly at the benzene ring and generates a benzyl radical/cyclohexadienyl radical pair. From here the reaction

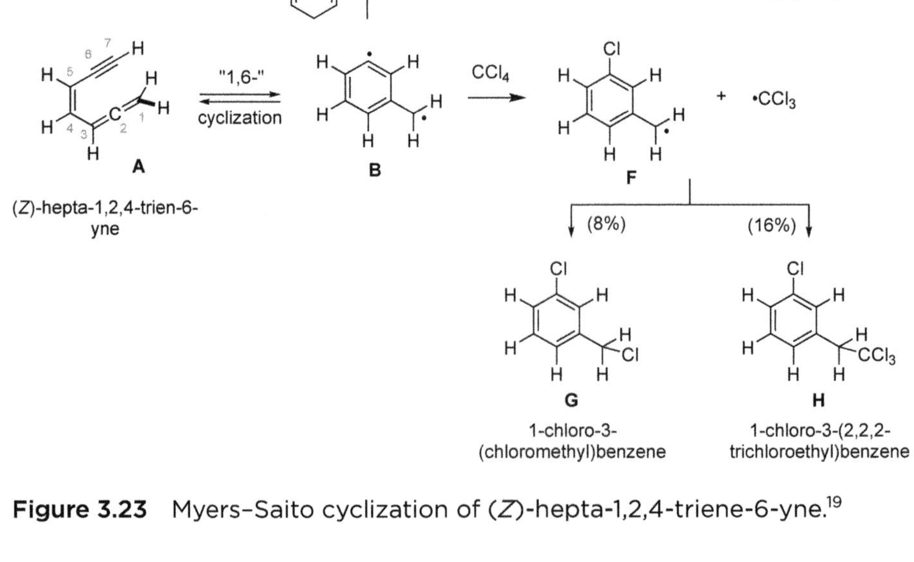

Figure 3.23 Myers–Saito cyclization of (Z)-hepta-1,2,4-triene-6-yne.[19]

continues either through a second hydrogen atom transfer step (to generate toluene (**D**) and benzene in 60% yield) or to the recombination product **E** (in the form of two diene regioisomers) in 40% yield. Trapping of biradical **B** by CCl_4 is slightly less effective in that the combined yields of isolatable products amounts to only 24%. Formation of these products can again be rationalized by initial reaction of the phenyl radical fragment in **B** to the 3-chlorobenzyl radical **F**. The reaction then continues through either a second chlorine atom transfer to dichlorotoluene **G** or the recombination product **H**.

Studying the influence of radical-stabilizing substituents on the course of the Myers–Saito cyclization, M. Schmittel and coworkers identified a second cyclization mode in substituents carrying substrates at the alkyne terminus of the reaction system. As shown for the phenyl-substituted substrate, as an example in Figure 3.24, six-membered ring formation through Myers–Saito cyclization competes with the five-membered ring-forming "Schmittel-cyclization" mode, in which one of the unpaired spins ends up on the phenyl-substituted alkyne terminus. Performing this thermally induced cyclization in cyclohexa-1,4-diene as the solvent yields only a small amount of Myers–Saito cyclization product **C** (3%) together with a larger amount of five-membered ring product **E** (as a mixture of stereoisomers, 11%).

The cyclization reactions of enediynes and eneyne allenes presented above play an important role in the anti-cancer activity of a number of potent and highly toxic natural products. An example for the enediyne class is dynemicin A, produced by the soil bacterium *Micromonospora chersina*. The purple blue color of this compound derives from its anthraquinone chromophore, but its cytotoxicity is due to

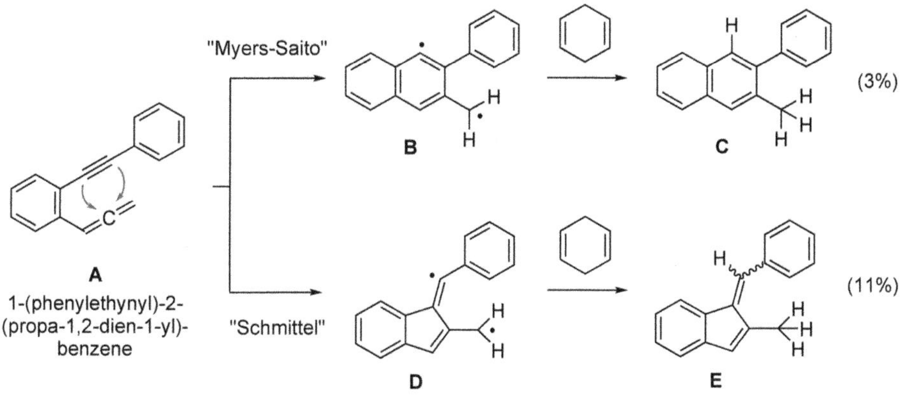

Figure 3.24   Competing Myers–Saito and Schmittel cyclizations in 1-(phenylethinyl)-2-(propa-1,2-dien-1-yl)benzene.[20]

the enediyne unit embedded in a ten-membered ring system (Figure 3.25). Binding of this compound to the DNA minor groove triggers reduction of its epoxide ring through glutathione (GST). This causes a small change in ring strain energy and is sufficient to trigger Bergman cyclization of the enediyne fragment. The cytotoxic properties of the bound drug then result from hydrogen atom abstraction from the DNA backbone through the resulting 1,4-diradical.

Neocarzinostatin is an anti-tumor antibiotic produced by *Streptomyces macromomyceticus* as a non-covalent conjugate of the highly reactive neocarzinostatin chromophore and its surrounding and stabilizing apoprotein. Binding of the neocarzinostatin chromophore to the DNA minor groove and addition of glutathione to the 1,3-diene-5-yne subsystem generate a highly reactive ten-membered eneyne allene ring system. Myers–Saito cyclization of this intermediate generates a biradical intermediate, whose hydrogen abstraction from the DNA backbone is again responsible for the cytotoxicity of neocarzinostatin.

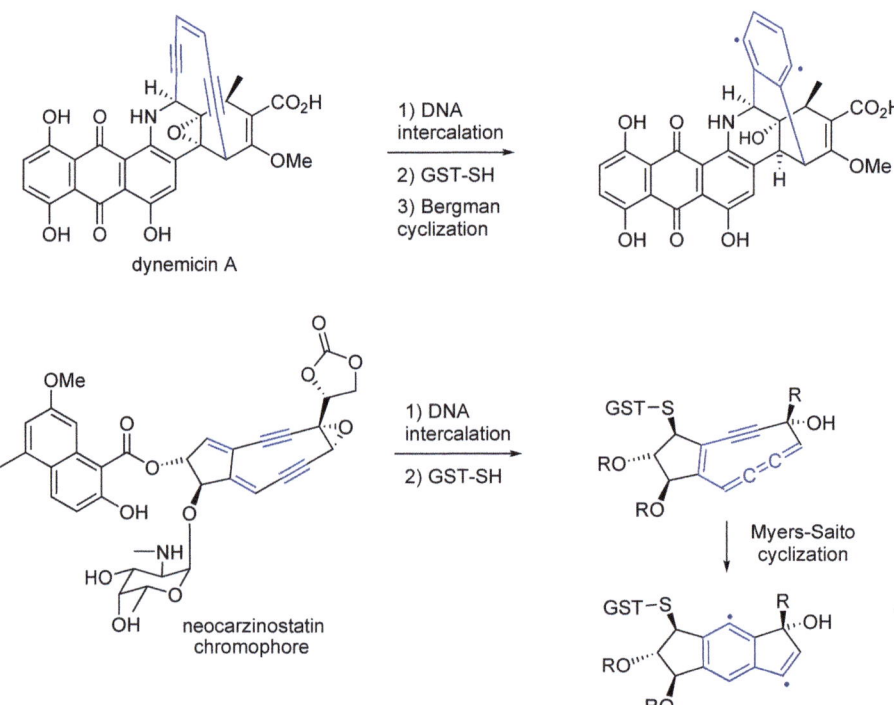

Figure 3.25 Structures and modes of action of dynamicin A and neocarcinostatin.

3.2 Cycloaddition Reactions

Cycloaddition reactions can be categorized by either the number of π-electrons or the size of the π-systems (number of atoms) actively involved in the bond-making processes (or both). For the two most common cycloaddition reactions, the (4+2) cycloaddition reaction (Diels–Alder reaction) and the 1,3-dipolar cycloaddition reaction (Huisgen reaction), this is exemplified in Figure 3.26. The azomethine ylide/ethylene cycloaddition shown in Figure 3.26a involves a π-system stretching across three centers in the ylide and an alkene with two centers, which classifies this reaction as a (3+2) cycloaddition in topological terms. The ylide π-system harbors four π-electrons, which, together with the two π-electrons in the alkene, classifies the reaction as a [4+2] cycloaddition in terms of the overall electron count. Please observe that the reaction topology is given in parentheses, while the overall electron count is given in square brackets. The Diels–Alder reaction shown in Figure 3.26b is of "4+2"-type for both the topology and the electron count, which may have led to the widespread practice of using square and round brackets interchangeably in the chemical literature.

Most of the synthetically useful cycloadditions generate two new σ-bonds in the process, which is also the case for the two examples shown in Figure 3.26. If the two σ-bonds are created in a single kinetic step the reaction is termed **"concerted"**, otherwise the reaction is **"stepwise"**. This mechanistic question can most easily be discussed with reference to the More O'Ferrall–Jencks diagram shown in Figure 3.27 for the example of the (4+2) cycloaddition reaction between buta-1,3-diene and ethylene. The two reactants are positioned in the lower left corner of the diagram, with the two axes defined as the bond order (b.o.) between the C1/C6 and C4/C5 centers involved in σ-bond

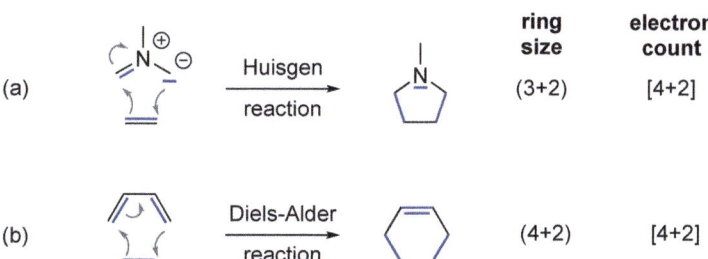

Figure 3.26 Prototypical examples for (a) the Huisgen reaction (1,3-dipolar cycloaddition), and (b) the Diels–Alder cycloaddition reaction.[21]

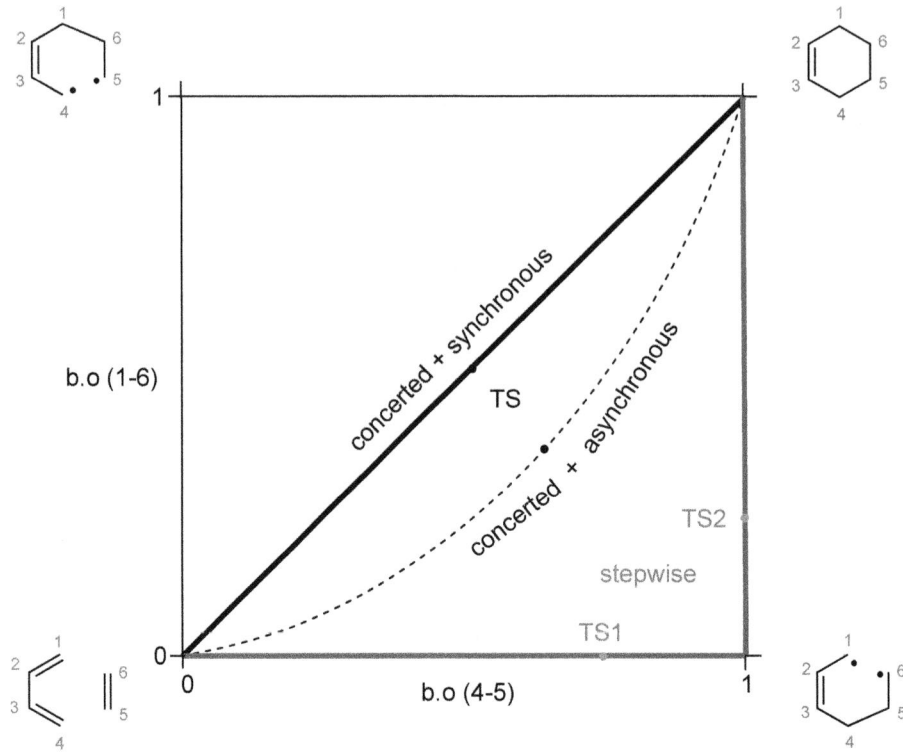

Figure 3.27   More O'Ferrall–Jencks diagram for the (4+2) cycloaddition between buta-1,3-diene and ethylene.

formation. The cycloaddition product with two newly formed σ-bonds is located in the upper right corner of the diagram, where both coordinates reach a value of 1.0. The other two corners of the diagram are occupied by (hypothetical) biradical intermediates, where only one of the two σ-bonds has formed. With these coordinate definitions, the concerted pathway leads in a single kinetic step from the lower left to the upper right corner. If the reaction is not only concerted, but also **synchronous**, the reaction pathway is identical to the diagonal of the diagram and the two σ-bonds that are forming have identical bond orders at each point along the reaction pathway. In Diels–Alder reactions involving donor/acceptor-substituted reaction partners, the reaction pathway may not follow this idealized line, but still proceeds in a single kinetic step (with only a single transition state). The reaction is then termed **asynchronous** (but still concerted). This differs from a stepwise process, where one of the σ-bonds is made first, followed by formation of a discrete biradical (or zwitterionic) intermediate.

Formation of the second σ-bond then occurs in a second, kinetically distinct reaction step (with an associated second transition state).

3.2.1 (2+2) Cycloaddition Reactions

The parent system for this type of cycloaddition reaction is the dimerization of two ethylene molecules to cyclobutane. Whether this reaction is symmetry allowed as a concerted [2+2] cycloaddition can be analyzed using an orbital correlation diagram closely similar to that used in Section 3.1 for electrocyclic reactions. The analysis starts by identifying a vertical and a horizontal mirror plane (σ_v and σ_h) as valid symmetry elements for reactants, products and all points along the reaction pathway, and thus as useful ordering criteria for the orbitals actively involved in bond making and breaking (Figure 3.28). While this is easy to see for the cyclobutane product, the π-orbitals of each individual ethylene unit are not symmetric or antisymmetric with respect to the vertical mirror plane σ_v. As shown in Figure 3.28 we therefore form symmetry-adapted combinations of

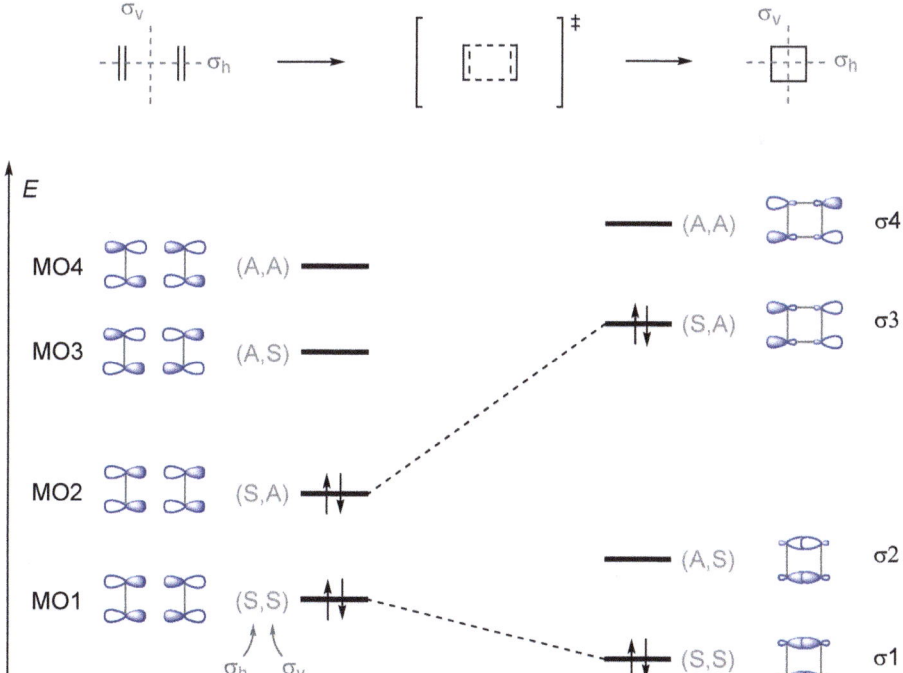

Figure 3.28 Orbital correlation diagram for the concerted [2+2] cycloaddition reaction of two ethylene molecules.

these π-orbitals that are symmetric or antisymmetric with respect to both mirror planes. The lowest energy combination, termed "MO1", is symmetric with respect to both (horizontal and vertical) mirror planes and is thus labeled as "(S,S)". The four π-type orbitals on the reactant side transform into four σ-type orbitals on the product side, the lower two corresponding to combinations of two σ(C–C) orbitals, and the upper two to combinations of the corresponding σ*(C–C) orbitals. Comparing the orbital symmetry labels on the reactant and product side we see that the lowest lying MOs on both sides are of "(S,S)" type. The two electrons placed in the lowest MO on the reactant side will thus end up in the lowest MO also on the product side. The next two electrons on the reactant side are placed in (S,A) symmetric MO2, which correlates in symmetry with the third highest MO on the product side, termed "σ3". Maintaining orbital symmetry along the reaction pathway will thus lead to the product in an electronically exited state, and the reaction is thus not allowed according to the Woodward–Hoffmann rules.

A similar result is obtained using the **frontier molecular orbital (FMO)** theory approach developed by K. Fukui. According to this model, pericyclic reactions are allowed if the HOMO of one reactant interacts favorably with the LUMO of the other reactant in the proposed transition-state geometry. Applying this idea to the [2+2] cycloaddition of two ethylene molecules first requires identification of the HOMO and LUMO orbitals of the unchanged reactants. As shown in Figure 3.29 the HOMO and LUMO orbitals of the two reacting ethylene units are identical, and it is therefore sufficient to analyze the interaction of the LUMO of the ethylene unit on the left with the HOMO of the ethylene unit on the right. The ethylene HOMO on the right has no phase change between the two carbon atoms, while the opposite is true for the ethylene LUMO on the left. Bringing these two orbitals face-to-face in the [2+2] cycloaddition transition state allows them to overlap favorably only at one side (in this case, the upper side), while no favorable overlap (shown as a broken grey arc in Figure 3.29) can be achieved on the other side. Thus, due to the incompatible structures of the frontier orbitals of ethylene, no stabilizing interactions can be generated in the transition state of the [2+2] cycloaddition reaction, and the reaction is thus not allowed according to FMO theory. This is at least true as long as we assume a transition-state structure involving the face-to-face orientation of the reacting ethylene molecules shown in Figure 3.29. Before discussing alternative transition-state structures, we will first introduce two new terms required to specify how two frontier orbitals can possibly

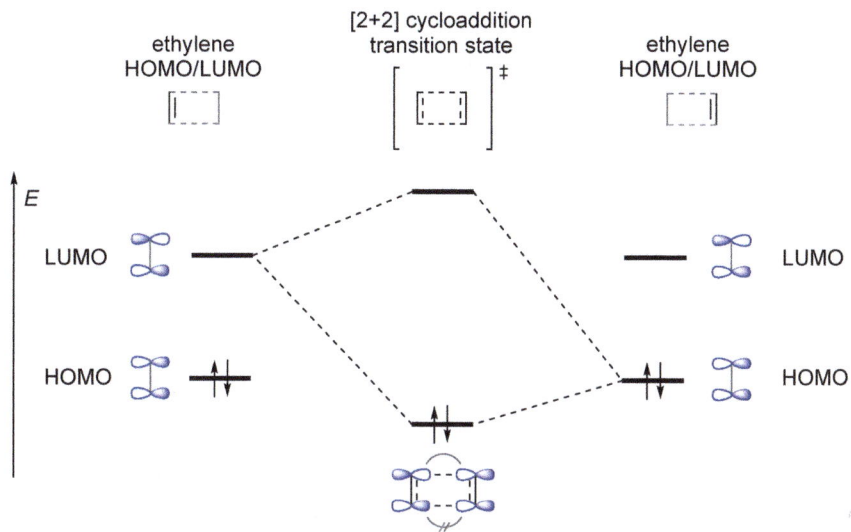

Figure 3.29 FMO analysis of the concerted [2+2] cycloaddition reaction of two ethylene molecules.

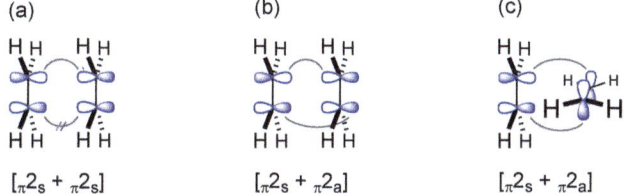

Figure 3.30 Possible FMO interactions in (a) the face-to-face [$_\pi 2_s + _\pi 2_s$]; the (b) face-to-face [$_\pi 2_s + _\pi 2_a$]; and (c) the side-on [$_\pi 2_s + _\pi 2_a$] orientations for the [2+2] cycloaddition reaction of two ethylene molecules.

interact with each other. The silent assumption made in the diagram in Figure 3.29 is that orbital interactions at both ends of the ethylene units involve the same side (or face) of the reacting π-systems. This type of interaction is shown again in Figure 3.30 and is termed "**suprafacial**". The interaction mode analyzed in Figure 3.29 is thus that of the [$_\pi 2_s + _\pi 2_s$] cycloaddition of two ethylene molecules, the "π" indicating reactions of π-systems, the "2" indicating the number of electrons involved in the respective π-system, and the subscript "s" indicating its suprafacial interaction in the bond-making process.

The term "**antarafacial**" is used to indicate bond formation at one end of the π-systems to one side and at the other end to the other side. HOMO/LUMO interactions in the face-to-face transition state for the dimerization of ethylene can, for example, be favorable if one of the components reacts suprafacially and the other component reacts antarafacially, as shown schematically in Figure 3.30b. For systems as small as ethylene it is, however, impossible for steric reasons to form one σ-bond on one side and the other σ-bond on the other side of the C–C double bond. Arranging the two reacting ethylene units in a side-on fashion, as shown in Figure 3.30c, and again assuming a $[_\pi 2_s + _\pi 2_a]$ process, does improve the chances for constructive overlap at all four reaction centers, but also suffers from repulsive interactions between the remaining σ(C–H) bond network. In conclusion we can state that the full orbital correlation diagram shown in Figure 3.28 as well as the FMO analysis shown in Figures 3.29 and 3.30 agree, in that the [2+2] cycloaddition of two ethylene molecules is an unfavorable process. This is in line with experimental observation that the dimerization of ethylene proceeds slowly even at elevated temperatures and most likely involves a stepwise reaction pathway.

Even for substituted alkenes (2+2) cycloaddition reactions are rarely observed, except for reactions of donor-substituted with acceptor-substituted alkenes. All the available information points, in this case, to a stepwise mechanism involving zwitterionic intermediates, as shown in Figure 3.31 for the example of (Z)-1-ethoxyprop-1-ene (**A**) with tetracyanoethylene (TCNE, **B**). These alkenes react to form zwitterionic adduct (Z)-**C**, where the formal positive charge is stabilized through the neighboring ethoxy group and the formal negative charge is stabilized through two cyano groups. Collapse of this zwitterionic adduct leads to the formal (2+2) cycloadduct (Z)-**D**, where the (Z)-stereochemistry of alkene **A** is conserved. Product analysis indicates the presence of small amounts of cycloadduct (E)-**D**, where the ethoxy and methyl substituents occupy opposite positions on the cyclobutane ring. (Z)- to (E)-isomerization can most easily be rationalized assuming C–C bond rotation at the stage of the zwitterionic intermediate **C**. In apolar organic solvents such as benzene the lifetimes of these intermediates are comparatively short and only little (Z/E)-isomerization is observed. More polar solvents stabilize the zwitterionic intermediates and thus increase the chances for their (Z/E)-isomerization. As a consequence a (Z/E) ratio of 85:15 is observed for the reaction of **A** and **B** in acetonitrile (Figure 3.31).

The (2+2) cycloaddition reaction shown in Figure 3.31 is one of the many mechanistic studies employing alkene stereochemistry as a "marker" property. The logic of analyzing this (2+2) cycloaddition reaction as a stepwise process rests on the assumption that (Z/E)-isomerization occurs exclusively at the stage of the zwitterionic

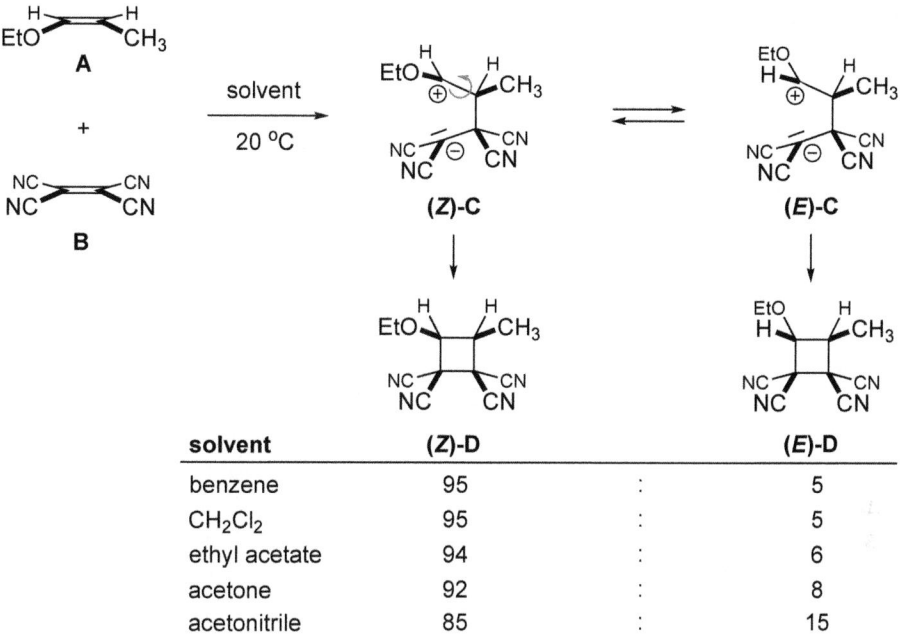

solvent	(Z)-D		(E)-D
benzene	95	:	5
CH$_2$Cl$_2$	95	:	5
ethyl acetate	94	:	6
acetone	92	:	8
acetonitrile	85	:	15

Figure 3.31 (2+2) cycloaddition of (Z)-1-ethoxyprop-1-ene (**A**) with tetra-cyanoethylene (TCNE, **B**) together with Z/E product ratios in different solvents.[22]

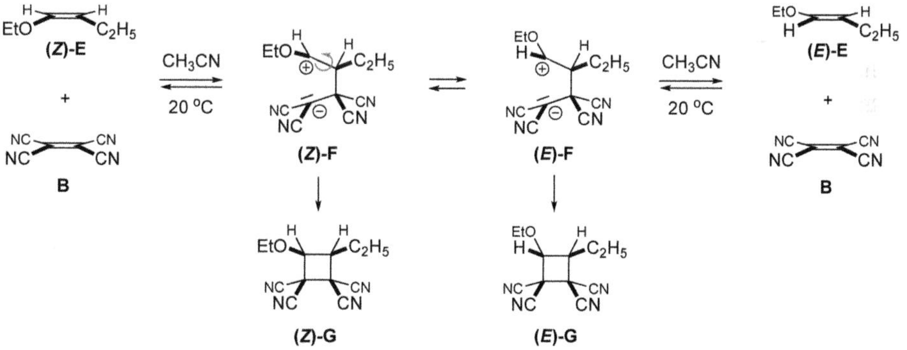

Figure 3.32 (2+2) cycloaddition of (Z)-1-ethoxybut-1-ene ((**Z**)-**E**) with tetracyanoethylene (TCNE, **B**) in acetonitrile.[22]

intermediates. This assumption is rarely tested, but has been carefully studied in the context of (2+2) cycloaddition reactions with TCNE (**B**) (Figure 3.32). Using (Z)-1-ethoxybut-1-ene ((**Z**)-**E**) as the substrate, Huisgen and Steiner indeed find that unreacted 1-ethoxybut-1-ene (**E**) contains appreciable amounts of the (E)-isomer, the exact amount depending on the reactant concentrations used. In this particular case detailed kinetic studies show that some (but not all) of the isomerized cycloadduct (*E*)-**G** derives from reaction of isomerized alkene (*E*)-**E**.

From a synthetic point of view (2+2) cycloaddition reactions involving electron-rich alkenes are particularly attractive. In the example shown in Figure 3.33 reaction between enamine **A** as the electron-rich alkene with acetylenedicarboxylate **B** as the electron-deficient reaction partner initially produces cycloadduct **C**. Conrotatory electrocyclic ring opening of the cyclobutene ring in **C** already proceeds rapidly at low temperature and generates the (*Z*,*E*)-cyclooctadiene **D** as the reaction product. Subsequent isomerization then yields the more stable (*Z*,*Z*)-cyclooctadiene **E** as the final product of this sequence.

A second class of important (2+2) cycloadditions are those of ketenes with alkenes or alkynes. Ketenes are so reactive that effective synthetic protocols usually combine their generation and subsequent trapping by (2+2) cycloaddition reactions in a one-pot procedure. A typical example of this approach is shown in Figure 3.34. Dichloroketene **B** is synthesized here through dehydrohalogenation of dichloroacetylchloride

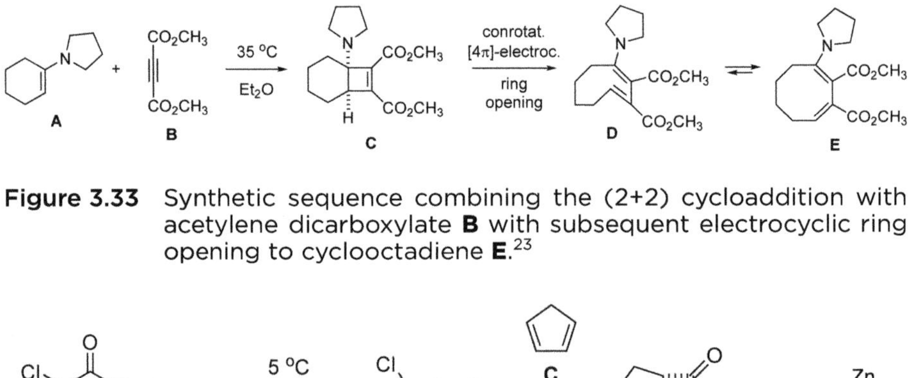

Figure 3.33 Synthetic sequence combining the (2+2) cycloaddition with acetylene dicarboxylate **B** with subsequent electrocyclic ring opening to cyclooctadiene **E**.[23]

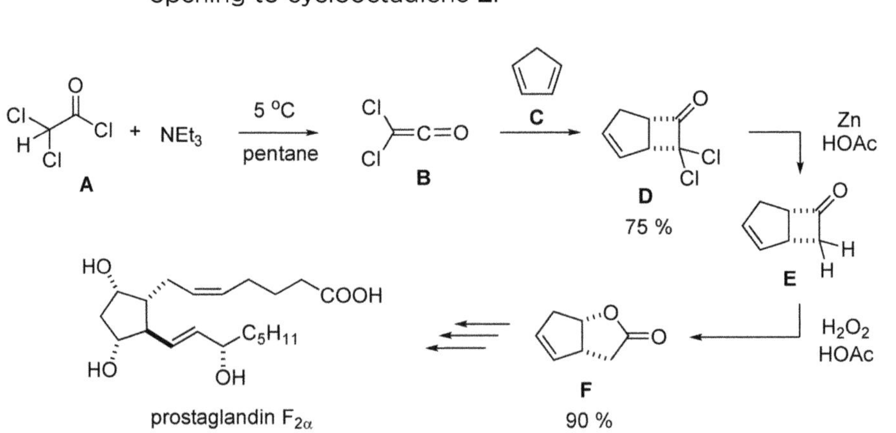

Figure 3.34 Racemic synthesis of prostaglandin $F_{2\alpha}$ employing the (2+2) cycloaddition of dichloroketene (**B**) with cyclopenta-1,3-diene (**C**) as the initial step.[24-26]

A with triethylamine. (2+2) Cycloaddition with cyclopenta-1,3-diene (**C**) yields 75% of adduct **D** as the only observed regioisomer. Dechlorination with zinc/acetic acid and subsequent Baeyer–Villiger oxidation with hydrogen peroxide/acetic acid (see Section 3.3.2 for mechanistic details of this reaction) then leads to bicyclic lactone **F** as a key intermediate of the synthetic sequence, which ultimately leads to prostaglandin $F_{2\alpha}$ in its racemic form.

As in the above example, ketenes react with dienes in a (2+2) rather than a (4+2) cycloaddition and also show a strong preference for reacting with their C–C rather than their C–O double bonds. One of the rare exceptions to this "**periselectivity**" (that is, the selectivity for one symmetry-allowed pericyclic reaction over another) is the reaction of 1,1-diphenylketene with cyclopenta-1,3-diene as the ketenophile (Figure 3.35). Following this reaction at low temperature by ^1H NMR spectroscopy, Yamabe and coworkers identified the (4+2) cycloadduct **D** as a true reaction intermediate, which then undergoes a [3,3]-sigmatropic rearrangement to (2+2) cycloadduct **E** as the final product. Performing the reaction at +20 °C yields no trace of intermediate **D** and delivers the (2+2) cycloadduct **E** directly. Re-analyzing the reaction at −20 °C and comparing various kinetic models, Singleton *et al.* concluded that product **E** is formed through at least two pathways, one being the interconversion between **D** and **E** and the other being the direct formation from **A** + **B**. The analysis of ^{13}C isotope effects and results from theoretical simulations suggest that the most likely scenario for this latter pathway involves a single transition state **C** for the formation of **D** and **E**. Transition state **C** can best be described as a charge-separated structure carrying a partial negative charge on the ketene moiety and in which C–C bond formation is far more advanced than C–O bond formation. This is also in line with substituent effects observed for a wide variety of ketene cycloaddition reactions, such as those shown in Figure 3.36 for 1,1-diphenylketene. For the reaction in

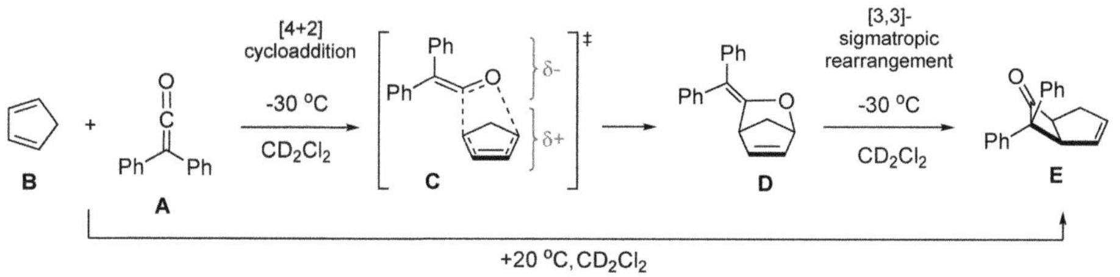

Figure 3.35 Sequential (4+2) cycloaddition and [3,3]-sigmatropic rearrangement of cyclopenta-1,3-diene (**B**) with 1,1-diphenylketene (**A**).[27,28]

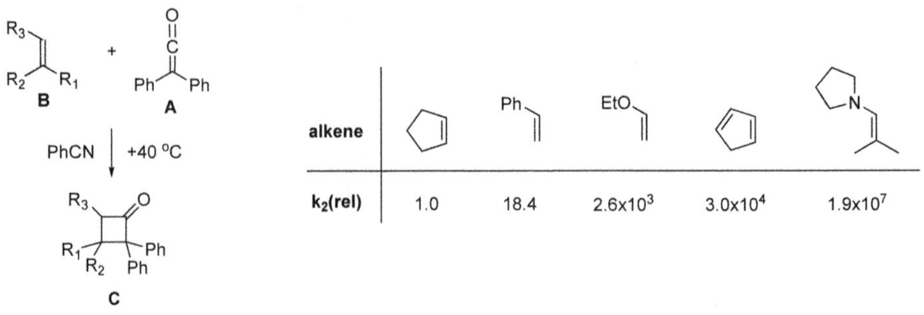

Figure 3.36 Substituent effects in (2+2) cycloaddition reactions with 1,1-diphenylketene (**A**).[29]

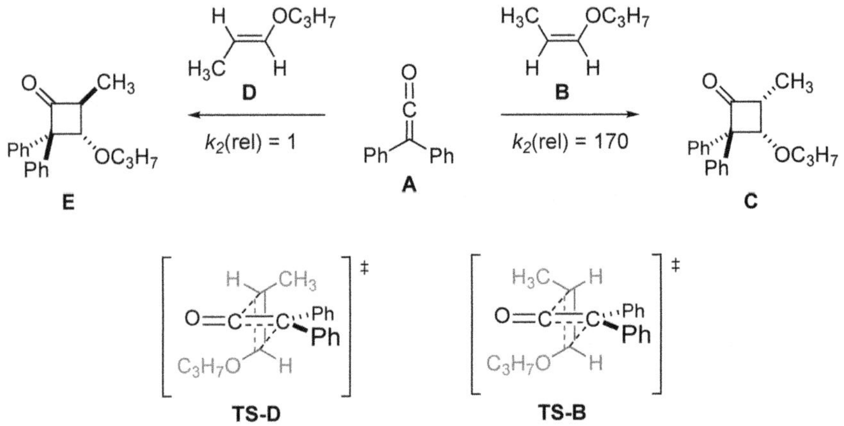

Figure 3.37 Stereocontrol in (2+2) cycloaddition reactions of 1,1-diphenylketene with (*E*)- and (*Z*)-enolethers.[30]

benzonitrile at 40 °C studied here, the (2+2) cycloadducts are the only products observed. The slowest reaction under these conditions is that for cyclopentene, with faster reactions being those with more electron-rich alkenes. The fastest ketenophile is 1-pyrrolidinoisobutene, which reacts approximately 1.9×10^7 times faster than cyclopentene. The enormous rate acceleration seen with more electron-rich alkenes in Figure 3.36 implies that 1,1-diphenylketene acts as the electrophile in these (2+2) cycloaddition reactions, which is fully in line with the charge distribution shown in the transition state in Figure 3.35.

Whether or not the reaction proceeds as a concerted cycloaddition reaction for the more electron-rich alkenes has been studied for 1,1-diphenylketene (**A**) in its reaction with enol ethers of (*E*) and (*Z*) configuration (Figure 3.37).

The reaction proceeds readily at room temperature with (*Z*)-1-propoxyprop-1-ene (**B**) to yield cycloadduct **C** with *cis* configuration as the only detectable product. The same reaction with (*E*)-1-propoxyprop-1-ene is significantly (170 times) slower and yields cycloadduct **E** with *trans* configuration as the only detectable product. The observed stereospecificity implies that the reaction is either fully concerted in both cases, or passes through short-lived intermediates, whose lifetimes are too small to allow for *cis/trans* isomerization. The different reaction rates can be understood assuming transition-state structures where the alkene attacks the ketene in a twisted side-on fashion. This particular structure results from orbital interaction considerations (see below). As shown in transition-state structure **TS-B** for the reaction of (*Z*)-1-propoxyprop-1-ene (**B**), this orientation avoids unfavorable steric interactions between the ketene phenyl groups and the two alkene substituents (Figure 3.37). This is not possible in transition state **TS-D** for reaction of the alkene with (*E*) configuration, where one of the two alkene substituents always comes into contact with the ketene phenyl groups.

How can the (2+2) cycloaddition be so facile and (and often stereospecific) in ketenes, but not in alkenes? Part of the answer can be found in the orbital structure of the ketene π-system (Figure 3.38a). This is actually composed of two subspaces oriented orthogonally relative to each other. The larger of the two is of allyl-type and harbors the π-electrons of the formal C–C double bond and two of the oxygen lone-pair electrons. The second of these allyl molecular orbitals

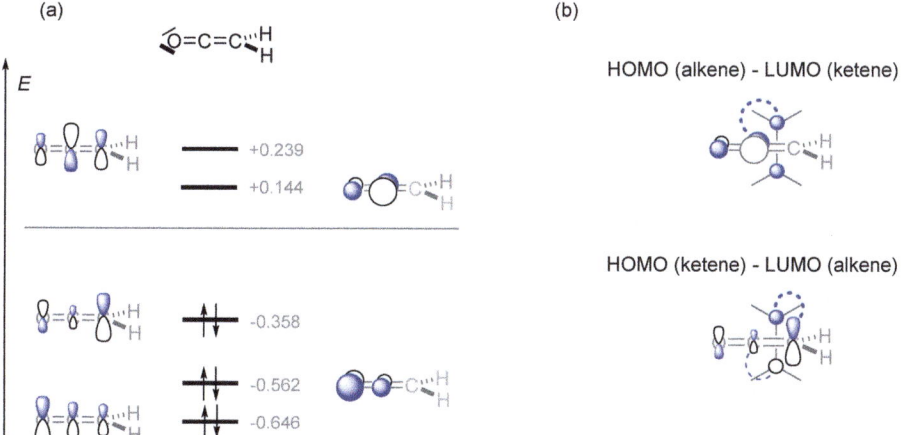

Figure 3.38   (a) Molecular orbitals of the ketene π-system (orbital energies are given in a.u.) and (b) important frontier orbital interaction in ketene (2+2) cycloaddition reactions.[31]

represents the overall HOMO of the system and has the largest coefficient at the terminal carbon atom. The second subsystem is essentially the π-system for a C–O double bond (and we neglect some mixing with the adjacent C–H bond orbitals at this point). The higher lying orbital of this second set has the largest coefficient at the central ketene carbon atom and represents the LUMO of the overall system.

All experimental and theoretical studies point to a (2+2) cycloaddition transition state where the alkene approaches the ketene C–C double bond in a twisted side-on orientation (Figure 3.38b). This points the ketene LUMO with its largest coefficient at the carbonyl carbon atom straight onto one of the alkene HOMO orbital positions in a way that is also observed for carbene or vinyl cation additions to alkenes. A second (much weaker) interaction exists between the second alkene carbon atom and the methylene carbon atom, which together satisfy the orbital symmetry conditions for allowed $[_\pi2_s + _\pi2_s]$ cycloaddition reactions. What makes this analysis unsatisfactory is the fact that the alternative HOMO (ketene)–LUMO (alkene) interaction is symmetry allowed only for the case of a $[_\pi2_s + _\pi2_a]$ process, the ketene HOMO structure being responsible for the antarafacial component. This has led to the widespread practice of ignoring the actual ketene LUMO and focusing on the (LUMO+1)–HOMO (alkene) interaction instead. This can easily be shown to give proper orbital interactions for a $[_\pi2_s + _\pi2_a]$ process, but leaves us with the question why the lower lying LUMO should not be considered. Part of the problem in ketene cycloadditions is the strongly polarized nature of both the ketene HOMO and LUMO, the former being localized mainly on the methylene carbon atom, and the latter (flipped by 90°) localized mainly on the carbonyl carbon atom. An alternative view of the ketene + alkene [2+2] cycloaddition reaction is therefore that of initial electrophilic attack of the ketene carbonyl atom at the alkene C–C double bond, which is guided by the LUMO (ketene)–HOMO (alkene) interaction alone. Subsequent collapse of the zwitterionic intermediate then involves the former ketene HOMO in its interaction with the second alkene center. Should the lifetime of the intermediate be negligible, then we will see the result of a concerted [2+2] cycloaddition reaction.

3.2.2 (4+2)/[4+2] Cycloaddition Reactions

Cycloaddition reactions of (4+2) topology and [4+2] electron count play an outstanding role in organic synthesis and formally transform three C–C double bonds in the reactants into two C–C single bonds and a new double bond in the product. The essentials of this reaction type are most often discussed using the reaction of buta-1,3-diene with ethylene, shown in Figure 3.39. Whether this reaction is allowed as a concerted cycloaddition according to the Woodward–Hoffmann rules can be analyzed using the same approach employed already for [2+2] cycloaddition reactions in Section 3.2.1. The active orbitals involved on the reactant side are the π-orbitals of buta-1,3-diene and of ethylene, which can be sorted by their relative orbital energies. The relevant symmetry element for this transformation is the vertical mirror plane (σ_v) running through the center of butadiene and ethylene (at least if, for the moment, we make the simplifying assumption that butadiene is present in a fully planar *cis* conformation). On the product side we group the C–C σ and σ* orbitals into sets that are symmetric (S) or antisymmetric (A) to the common vertical mirror plane. To the resulting set of four σ and σ* orbitals we add the π and π* orbitals of the product C–C double bond in order to complete the set of active

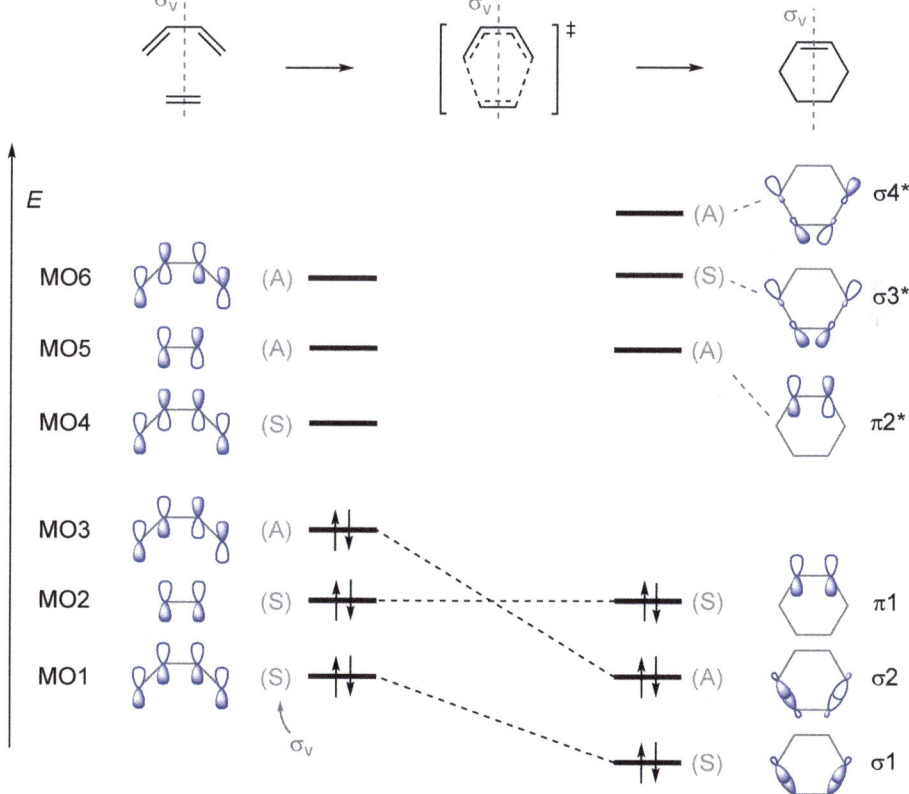

Figure 3.39 Orbital correlation diagram for the concerted (4+2)/[4+2] cycloaddition reaction of buta-1,3-diene with ethylene.

product orbitals. Following the electrons located in the lower three reactant orbitals to the product side, we note that the product is generated in its electronic ground state, and the concerted [4+2] cycloaddition of buta-1,3-diene with ethylene is thus allowed according to the Woodward–Hofmann rules.

This analysis, based on a very qualitative form of molecular orbital theory, compares well with results from best current theoretical studies and experimental mechanistic analysis. The most relevant experimental study employs deuterated buta-1,3-diene and ethylene substrates in order to follow the stereochemical course of the reaction (Figure 3.40). This particular (4+2) cycloaddition requires high temperature/high pressure conditions for acceptable reaction rates, and yields a mixture of deuterated cyclohexenes as products. Quantitative analysis of the product mixture involves ^{1}H NMR measurements on the epoxides formed from the original products through reaction with *meta*-chloroperbenzoic acid (mCPBA). In the limits of this type of analysis, reactions of (Z)- and (E)-dideuterioethylene yield cyclohexene products with full retention of configuration, which requires that the barrier for the concerted [4+2] cycloaddition is at least 15.5 kJ mol^{-1} lower than for any alternative stepwise process allowing for E/Z-isomerization through, for example, biradical intermediates.

Quantum chemical studies of the buta-1,3-diene + ethylene system confirm this picture (Figure 3.41): the concerted pathway through a transition state, in which both of the forming C–C single bonds have identical lengths of around 225 pm, is located around 92 kJ mol^{-1} above the separate reactants. Please observe that in this case reference is made to buta-1,3-diene in its more stable *trans*-conformation. A pathway for the stepwise (4+2) cycloaddition exists at higher energies and

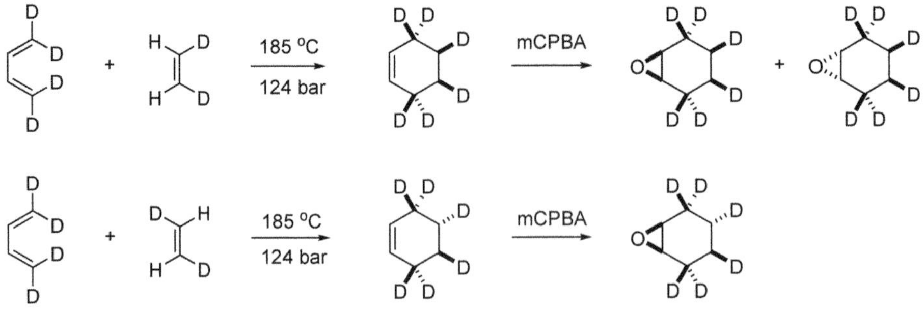

Figure 3.40 Experimental study of the (4+2) cycloaddition reaction of 1,1,4,4-tetradeuteriobuta-1,3-diene with (Z)- and (E)-1,2-dideuterioethylene.[32]

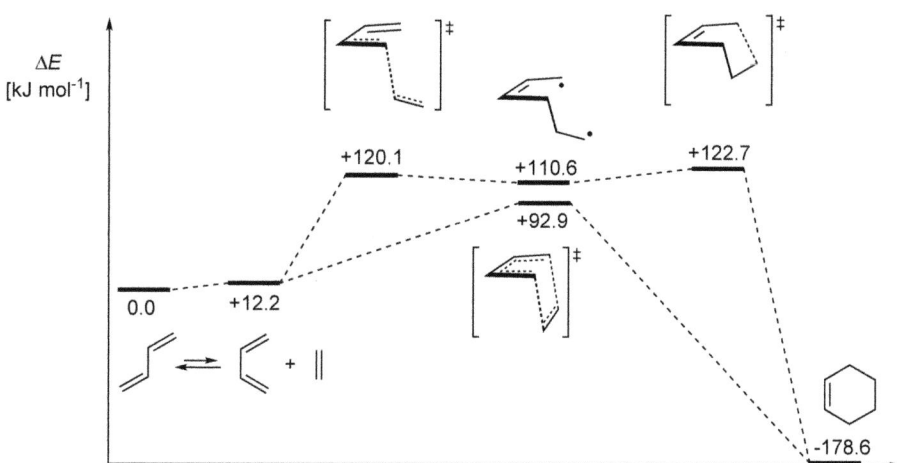

Figure 3.41 Concerted and stepwise pathways for the (4+2) cycloaddition reaction of buta-1,3-diene with ethylene (data in part from ref. 33).

involves initial formation of only one of the C–C single bonds. After passing through a biradical intermediate located in a shallow energy well, the second C–C single bond is made in a second step. The first of the two transition states in this stepwise pathway is located at least 27 kJ mol^{-1} higher than the transition state for the concerted process, in agreement with the interpretation of deuterium labeling experiments.

Analyzing the (4+2) cycloaddition of buta-1,3-diene with ethylene as a concerted $[_\pi 4_s + _\pi 2_s]$ process using the FMO approach reveals two important HOMO/LUMO interactions in the transition state (Figure 3.42). The first involves interaction of the HOMO of buta-1,3-diene with the LUMO of ethylene. Assembling these orbitals in the face-to-face geometry assumed for a concerted process shows that the terminal orbital positions involved in the C–C bond-forming process interact favorably. The same result is obtained for the interaction of the LUMO of buta-1,3-diene and the HOMO of ethylene, and we can thus conclude that the concerted $[_\pi 4_s + _\pi 2_s]$ cycloaddition of buta-1,3-diene and ethylene is a symmetry-allowed process.

The FMO analysis described above for the parent buta-1,3-diene + ethylene system is particularly helpful to rationalize substituent effects in Diels–Alder reactions. The majority of experimentally known systems fall into two categories (Figure 3.43).

1. The "normal" Diels–Alder reaction of electron-rich dienes, with comparatively high lying HOMO orbitals, with electron-poor alkenes (or more generally: **dienophiles**), with comparatively low

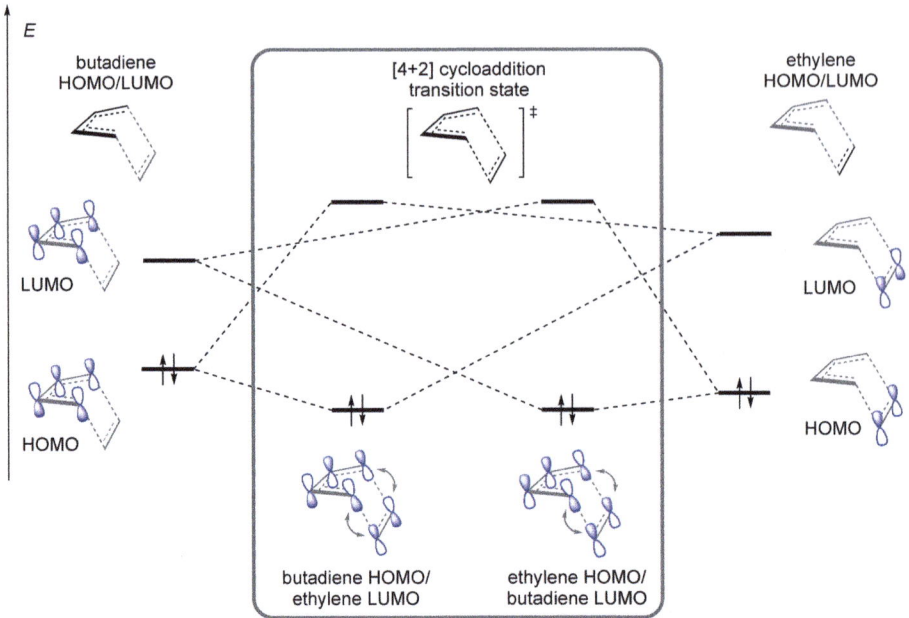

Figure 3.42 FMO analysis of the (4+2) cycloaddition of buta-1,3-diene with ethylene as a $[_\pi 4_s + _\pi 2_s]$ process assuming a face-to-face orientation of the two reaction partners.

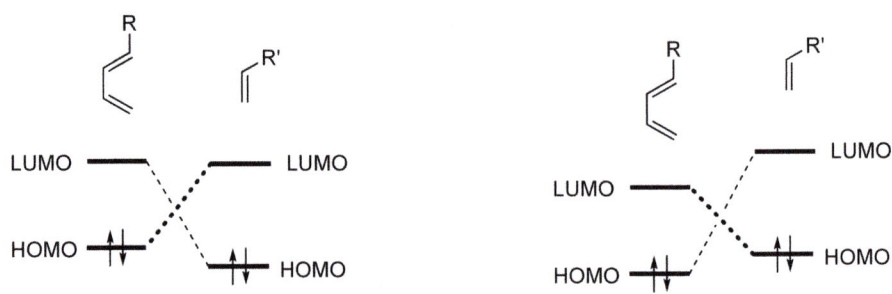

Figure 3.43 FMO interactions in $[_\pi 4_s + _\pi 2_s]$ cycloaddition reactions with (a) "normal" and (b) "inverse" electron demand.

lying LUMO orbitals. In this case the HOMO(diene)/LUMO(dienophile) interaction is dominant.

2. In the **"inverse electron demand"** Diels–Alder reaction the roles are reversed and electron-poor dienes react with electron-rich dienophiles. In this case the HOMO(dienophile)/LUMO(diene) interaction is dominant.

A third reaction category is conceivable where the substitution pattern of diene and dienophile is such that both orbital interactions shown in Figure 3.42 are similarly relevant. However, these latter reactions often proceed with low reaction rate and are therefore synthetically less relevant.

An impressive example of the enormous role of substituent effects in Diels–Alder reactions can be found in reaction rates for the reaction of cyclopenta-1,3-diene (CP) with cyano-substituted alkenes at 20 °C in dioxane as the solvent (Figure 3.44a). Reaction with acrylonitrile as the slowest reacting alkene proceeds at a rate that is already fast enough to be synthetically useful. Addition of a second cyano substituent lowers the energy of the alkene LUMO and thus accelerates the cycloaddition reaction. The effect is particularly large in 1,1-dicyanoethylene, which reacts $455\,000/10.4 = 43\,750$ times faster compared to acrylonitrile. Addition of further cyano substituents accelerates the reaction further, and the fastest reaction is found for tetracyanoethylene (TCNE), which shows a speeding up relative to acrylonitrile of more than seven orders of magnitude! A synthetically even more interesting strategy for increasing reaction rates in "normal" Diels–Alder reactions involves Lewis acid coordination of the dienophile. As shown in the example in Figure 3.44b, this works particularly well with dienophiles carrying carbonyl or ester substituents, whose complexation

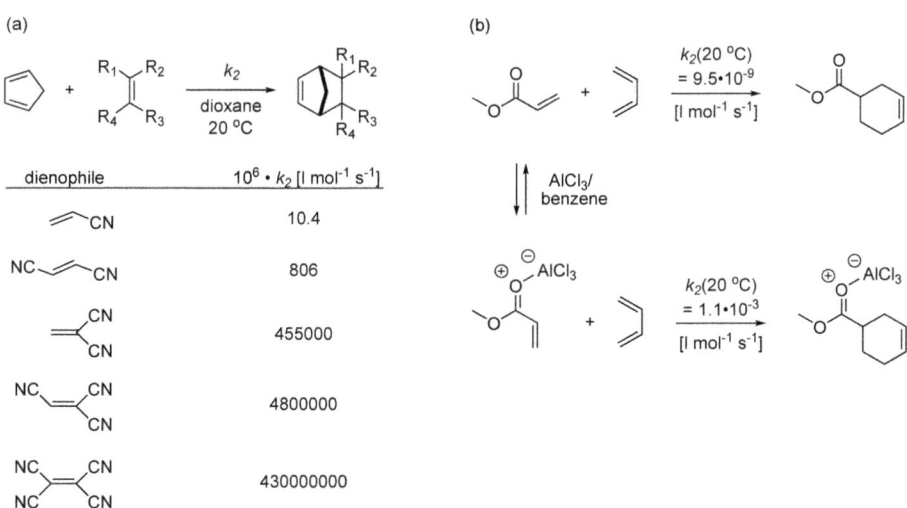

Figure 3.44 Rate acceleration of the "normal" Diels–Alder reaction through (a) introduction of electron-withdrawing substituents into the dienophile and (b) Lewis acid complexation of the dienophile.[34,35]

with Lewis acids, such as aluminium(III) chloride, lowers the LUMO energy substantially and leads, for the example of the methyl acrylate reaction with buta-1,3-diene, to a rate acceleration of five orders of magnitude under otherwise comparable conditions. Arrhenius analysis of reaction rates measured at different temperatures shows that the acceleration is mainly due to a reduction in activation energy (E_a for the uncatalyzed *vs.* catalyzed reactions amount to +73.3 *vs.* +43.5 kJ mol^{-1}).

Most of the inverse electron demand Diels–Alder reactions involve dienes where at least one of the carbon atoms has been replaced by more electronegative elements, such as nitrogen or oxygen, and thus belong to the class of **hetero-Diels–Alder** reactions. As discussed in Chapter 1, Section 1.3.3, this replacement lowers the energy levels of all diene π-type molecular orbitals, and, together with electron-withdrawing substituents, provides (hetero)dienes suitable for reaction with electron-rich dienophiles. A typical example for this scenario is shown in Figure 3.45a, where *N*-phenyl-1-aza-2-cyanobutadiene reacts as an electron-poor diene with dihydrofuran as the electron-rich dienophile to give the (4+2) cycloadduct in 87% yield.

A class of hetero-Diels–Alder reactions that work particularly well involve 1,2,4,5-tetrazines as the diene component. Hetero-Diels–Alder reactions with electron-rich alkenes, such as vinyl ethers, already

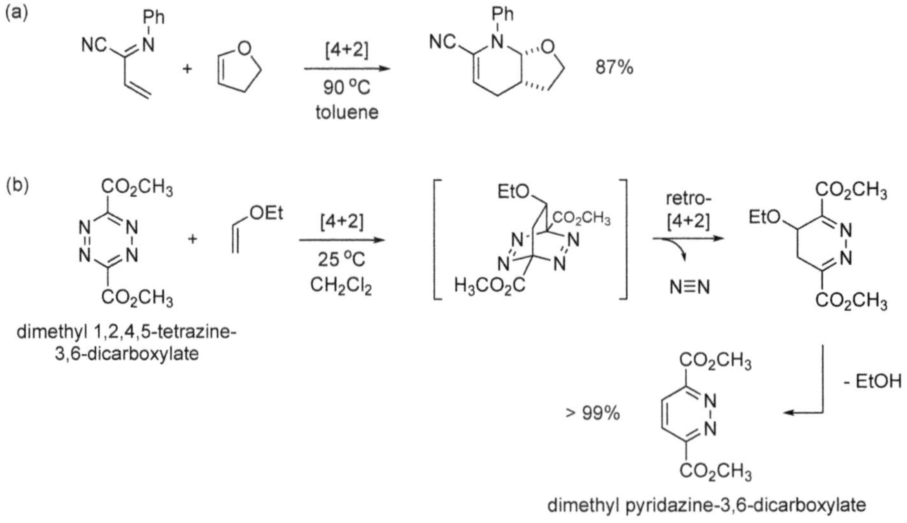

Figure 3.45 Inverse electron demand Diels–Alder reactions involving (a) 1-aza-butadienes and (b) 1,2,4,5-tetrazines as the diene component.[36,37]

proceed quite rapidly at room temperature for tetrazines carrying additional electron-withdrawing substituents, such as dimethyl 1,2 ,4,5-tetrazine-3,6-dicarboxylate (Figure 3.45b). The initially formed cycloadducts are not usually stable, but eliminate N_2 in a follow-up **retro-Diels–Alder reaction** (that is, a Diels–Alder reaction running backwards) to yield substituted dihydropyridazines as the first isolable product. In cases where the dienophile carries substituents easily eliminated as a nucleofuge, the initially formed dihydropyridazines react further to yield substituted pyridazines as the final products. In the example shown in Figure 3.45b the elimination of ethanol is indeed quite fast at room temperature, and dimethyl pyridazine-3,6-dicarboxylate is therefore isolated in near quantitative yield as the final product.

Diels–Alder reactions are not only influenced by electronic, but also by steric substituent effects. This also includes the effort for the diene substrate to assume the *cis* conformation required in the concerted [4+2] cycloaddition process. In cyclic dienes, such as cyclopenta-1,3-diene, the termini of the diene system find themselves already in the positions required for the [4+2] transition state, which leads to high reaction rates with dienophiles such as maleic anhydride (Figure 3.46). For other ring sizes, or for acyclic dienes, reaction rates are much lower, in comparison. A particularly dramatic substituent effect can be seen in the reactions of (*E*)- and (*Z*)-penta-1,3-diene, where the former reacts approximately five orders of magnitude faster than the latter. This is nothing else but a reflection of unfavorable

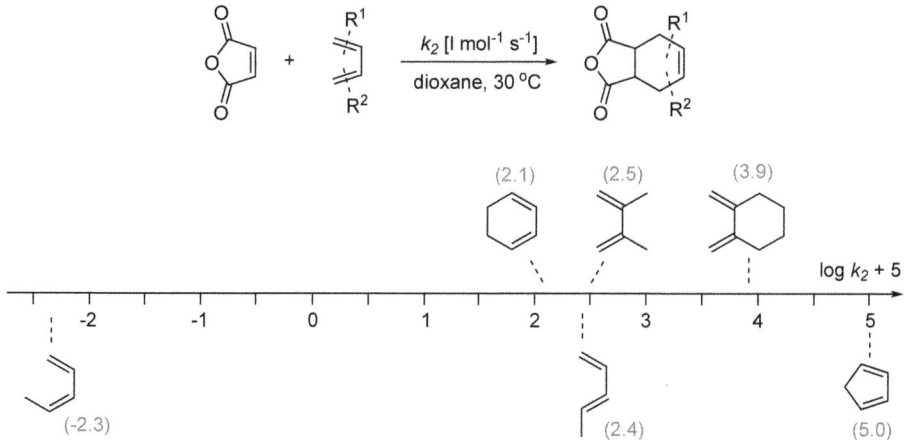

Figure 3.46 Reaction rates for the Diels–Alder reaction of maleic anhydride with selected dienes.[38]

syn-pentane interactions (see Chapter 1, Section 1.6.1) in the Diels–Alder transition state for (*Z*)-penta-1,3-diene, which requires the diene in its *cis* conformation.

Diels–Alder reactions of reactants with a non-symmetrical substitution pattern yield a mixture of regioisomeric cycloadducts. The regioselectivity is often low, as can be seen from the example of isoprene with acrylonitrile in Figure 3.47a. The two products formed here in a ratio of 75 : 25 carry the ring substituents either in 1,4- or 1,3-position. In analogy to the nomenclature of substituted benzenes, these are sometimes referred to as the "*para*" and "*meta*" products. A higher regioselectivity (and also a higher reaction rate) is observed under the same conditions for the reaction of isoprene with 1,1-dicyanoethylene, where the *para* : *meta* product ratio is now 91 : 9. Both reactions belong to the class of Diels–Alder reactions with normal electron demand, as is easily seen from the HOMO/LUMO orbital energies in Figure 3.47b.

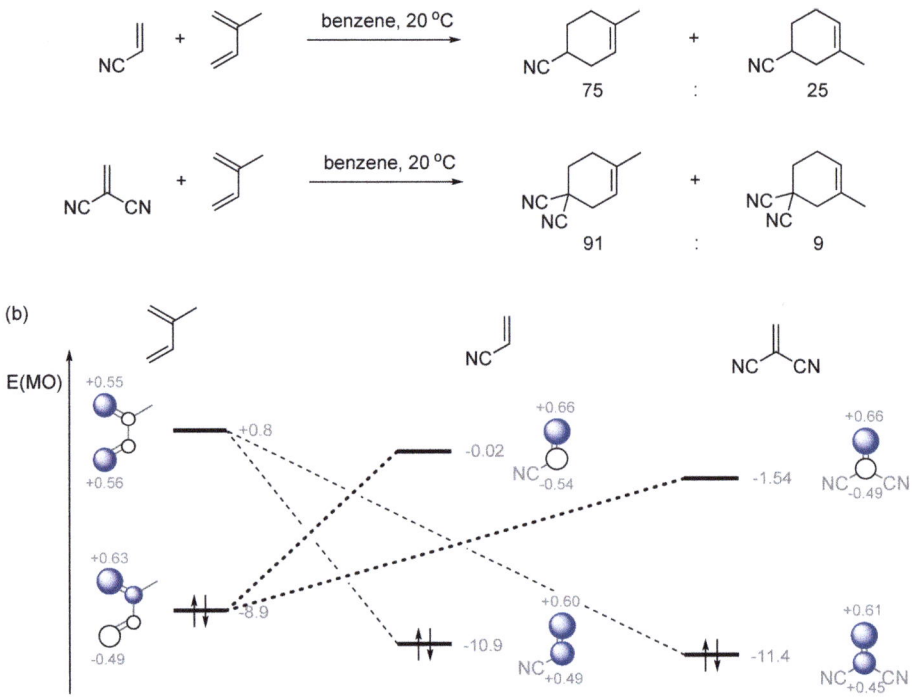

Figure 3.47   (a) Regioselectivity in the reaction of isoprene with acrylonitrile and 1,1-dicyanoethylene. (b) Orbital energies (in eV) and FMO coefficients for isoprene, acrylonitrile, and 1,1-dicyanoethylene.[39,40]

For acrylonitrile the HOMO(isoprene)/LUMO(acrylonitrile) energy gap amounts to $8.9 - 0.02 = 8.88$ eV, while a larger gap of $10.9 + 0.8 = 11.7$ eV exists between the HOMO(acrylonitrile) and the LUMO(isoprene). The HOMO(isoprene)/LUMO(dienophile) gap is even smaller for 1,1-dicyanoethylene at 7.36 eV, which is in agreement with the generally higher reaction rates observed for this alkene relative to acrylonitrile. Closer inspection of the structure of the isoprene HOMO shows that the terminus closer to the methyl substituent carries a slightly larger MO coefficient (+0.63) compared to the other diene terminus (−0.49). The same analysis of the acrylonitrile LUMO shows polarization towards the less substituted end of the C–C double bond. In the (energetically preferred) transition state leading to the *para* product the diene/dienophile termini with the largest FMO coefficients bind together. The size of this effect depends on how different the MO coefficients are in the respective FMO (here the dienophile LUMO), which also explains why reaction of 1,1-dicyanoethylene is more selective than reaction of acrylonitrile. The crude qualitative FMO analysis outlined above can be performed in a much more sophisticated, semiquantitative manner. But even then the accurate prediction of regioselectivities of Diels–Alder reactions remains a challenge. This is simply due to the rather small differences in activation barriers for formation of the two regioisomeric reaction products. Assuming that the Eyring equation (see Chapter 2, Section 2.1) provides a valid relationship between reaction rate constant k and activation free energy $\Delta G^{\ddagger}$, the activation free energy difference between *para* and *meta* product formation is given as $\Delta\Delta G^{\ddagger} = -RT \ln(k_{para}/k_{meta})$. For the moderately selective isoprene + acrylonitrile reaction with $k_{para}/k_{meta} = 75/25$, this implies a free energy difference between the respective transition states of only $\Delta\Delta G^{\ddagger} = 2.7$ kJ mol^{-1}. Most semiquantitative reactivity models in chemistry are simply not accurate enough to predict such small energy differences with confidence.

Diels–Alder reactions involving cyclic dienes show a stereochemical phenomenon often referred to as the "Alder rule" or "**Alder *endo* rule**". The cycloaddition products formed in these reactions are bicyclic in nature and thus feature two bridgehead (carbon) atoms and three bridge units of different lengths. If a substituent located on one of the bridge units points to the larger of the remaining two bridging units, its orientation is referred to as "*endo*", and the opposite orientation is termed "*exo*". In the example shown in Figure 3.48a, the uncatalyzed reaction of cyclopenta-1,3-diene with methyl acrylate yields a 78:22 ratio of the *endo/exo* cycloaddition products. Addition of aluminium(III) chloride accelerates the reaction significantly and also leads

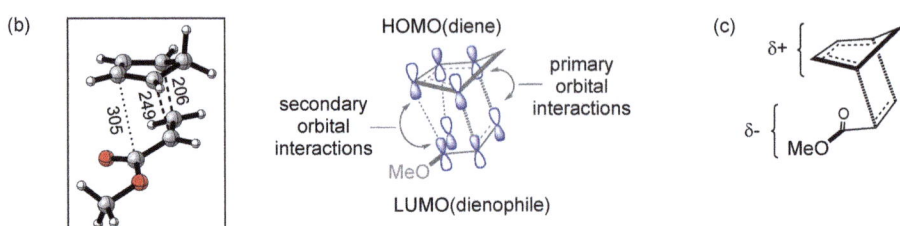

(b)   HOMO(diene)   (c)

Figure 3.48 (a) *Endo/exo* product ratios in the uncatalyzed and AlCl₃-catalyzed reactions of cyclopenta-1,3-diene with methyl acrylate. (b) Conceptual models for the "*endo* rule" together with (c) a three-dimensional rendition of the *endo* transition state. Selected distances are given in pm.[41]

to a much higher *endo/exo* product ratio of 95 : 5. In most cases it can be shown, through equilibration experiments, that the *endo* product is the less stable of the two cycloadducts, and any preference for *endo* cycloadducts is thus a kinetic phenomenon. The preferred formation of *endo* products has been rationalized with reference to a number of qualitative models. The most common model builds on specific structural features of *endo* transition states, such as the one obtained from quantum mechanical calculations in Figure 3.48b. The proximity of the reacting π-systems suggests that, in addition to the primary HOMO(diene)/LUMO(dienophile) interactions responsible for C–C bond formation, a second (weaker) set of interactions exist termed **"secondary orbital interactions"** (SOI). That these will materialize only in *endo* transition states is easily seen from Figure 3.48b, where the HOMO(diene) and LUMO(acrylate) orbitals are shown again to match their phases at both positions where single bond formation occurs. Further inspection of the two orbitals shows that phases also match between the centers not involved in actual bond making (C2/C3 of the diene; C–O double bond of acrylate). In the respective *exo* transition state these centers will be too far apart to interact constructively, and SOI effects are thus the sole property of *endo* transition states. The fact that AlCl₃ catalysis increases the *endo/exo* selectivity is, in the framework of the SOI model, a consequence of lowering the LUMO energy and polarization of the LUMO such that HOMO/LUMO interactions are stronger compared to the uncatalyzed case. A second qualitative model rationalizing the preferred formation of *endo* cycloadducts is based on electrostatic interactions between the reaction partners in the transition states. In Diels–Alder reactions of normal electron demand the (electron-rich) diene donates some electron density to the (electron-poor) dienophile in the transition state. This often leads to the charge distribution shown schematically in Figure 3.48c, where the *endo* transition state is more stable than the *exo* structure simply because of the more favorable relative orientation of the centers carrying positive and negative partial charges.

3.2.3 (3+2)/[4+2] Cycloaddition Reactions

Cycloaddition reactions of (3+2) topology and [4+2] electron count represent the second large class of synthetically important cycloaddition reactions and are often referred to as **Huisgen reactions** (or **1,3-dipolar cycloaddition reactions**). These derive from Diels–Alder reactions through formal replacement of the diene component by dipolar molecules as the 4π-electron component. Two classes of dipoles can be differentiated based on the geometry of the 3-center/4-electron component (Figure 3.49): (a) dipoles of propargyl/allenyl-type with linear geometry; and (b) allyl-type dipoles with bent geometry.

In Figure 3.49, dipoles of both types are ordered from top to bottom by their approximate LUMO energies, which apparently decrease with an increasing number of electronegative atoms. The most electron-poor dipole (by far) is ozone. The 1,3-dipoles shown in Figure 3.49 react with alkenes or alkynes as reaction partners (then termed **dipolarophiles**) in a symmetry-allowed, concerted process to five-membered ring cycloadducts, and the Huisgen reaction thus represents a very attractive and practical way for the synthesis of five-membered ring heterocycles. The fact that these reactions are symmetry allowed as concerted [4+2] cycloadditions follows from the analysis of FMO interactions in the cycloaddition transition state, as shown in Figure 3.50. Here we find, on the left side, the three molecular orbitals of the allyl π-system of the dipole **abc**, which is isoelectronic with the allyl anion. For the dipolarophile, designated **xy**, we consider the two π-MOs of the x–y double bond. As already seen for Diels–Alder reactions, phase-matching is favorable for the combinations HOMO(dipole)/LUMO (dipolarophile) and HOMO(dipolarophile)/LUMO(dipole). This result is also obtained when using the FMOs of propargyl-type dipoles. Note

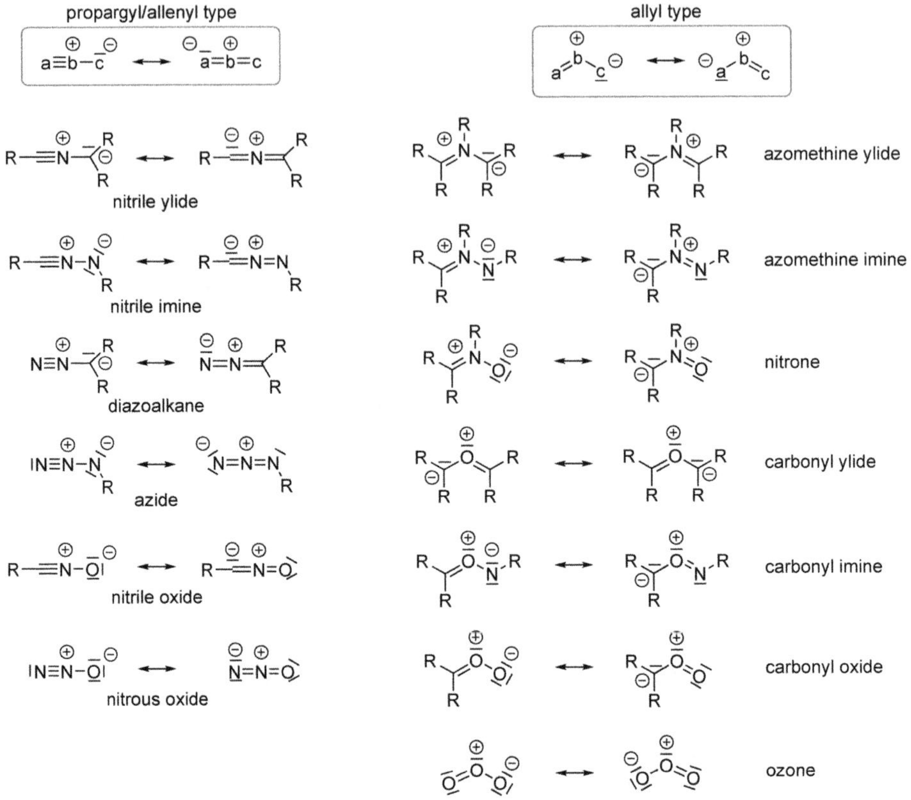

Figure 3.49 Selected dipoles used in Huisgen reactions.

that the "HOMO/LUMO" terminology used here refers to the analysis solely focused on the reacting π-type molecular orbitals. Given the large number of lone-pair electrons in all dipoles shown in Figure 3.49, the true HOMO of the system may or may not correspond to the orbitals shown in Figure 3.50.

Depending on the FMO energies of dipoles and dipolarophiles, three limiting situations are often discussed for the Huisgen reaction (Figure 3.50). In **type I** Huisgen reactions electron-rich 1,3-dipoles react with electron-poor dipolarophiles, and the HOMO(dipole)/LUMO(dipolarophile) interaction is thus the dominating orbital interaction. In **type II** Huisgen reactions the HOMO(dipole)/LUMO(dipolarophile) and HOMO(dipolarophile)/LUMO(dipole) interactions are both of comparable importance. **Type III** Huisgen reactions are dominated by the HOMO(dipolarophile)/LUMO(dipole) interaction and involve the reaction of electron-poor dipoles with electron-rich dipolarophiles. Reaction rates for type I Huisgen reactions depend strongly on the

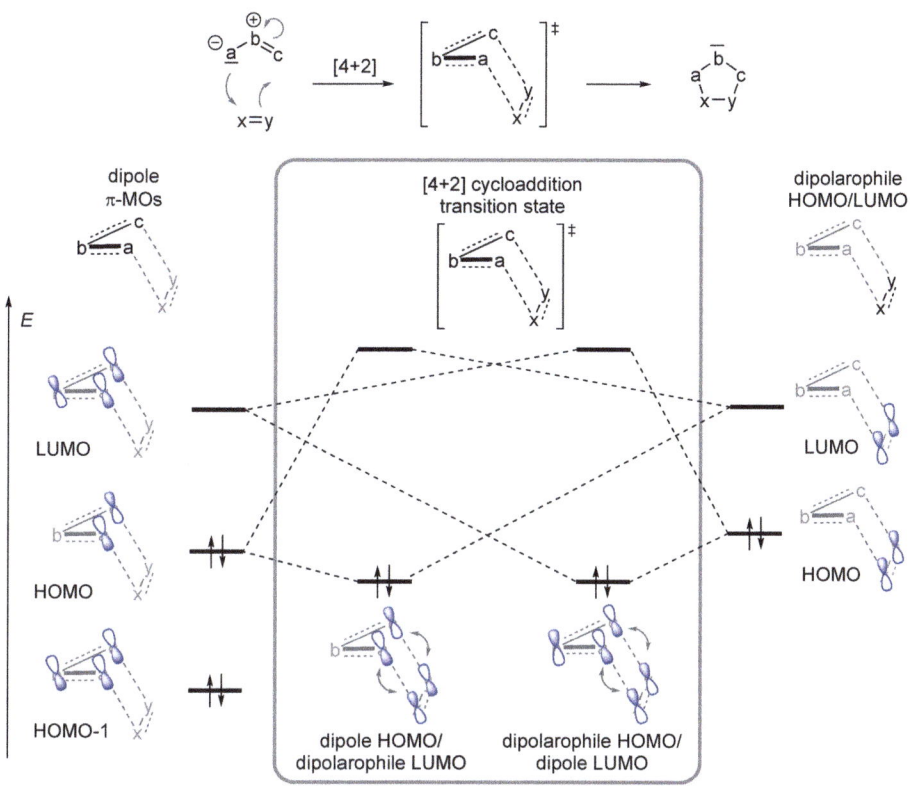

Figure 3.50 FMO analysis of the (3+2) cycloaddition of dipole **abc** with dipolarophile **xy** as a $[_\pi 4_s + _\pi 2_s]$ process assuming a face-to-face orientation of the two reaction partners.

substitution pattern of the dipolarophile. A typical dipole of this class is diazomethane, whose reaction with electron-deficient alkenes, such as ethyl acrylate, is seven (!) orders of magnitude faster than reaction with electron-rich alkenes, such as butyl vinyl ether (Figure 3.51a). Intermediate reactivities are observed for ethylene as the unsubstituted reference system, or for styrene as a typical alkene carrying a large π-substituent. This reactivity pattern is easily understood with respect to the dominating HOMO(diazomethane)/LUMO(dipolarophile) interaction, where electron-withdrawing alkene substituents lead to lower LUMO energies and thus more favorable orbital interactions. The HOMO of diazomethane is significantly larger at the carbon terminus as opposed to the nitrogen terminus. For dipolarophiles with equally polarized LUMO (*e.g.* acceptor-substituted alkenes) high regioselectivities are observed with preference for the regioisomer shown in Figure 3.51a.

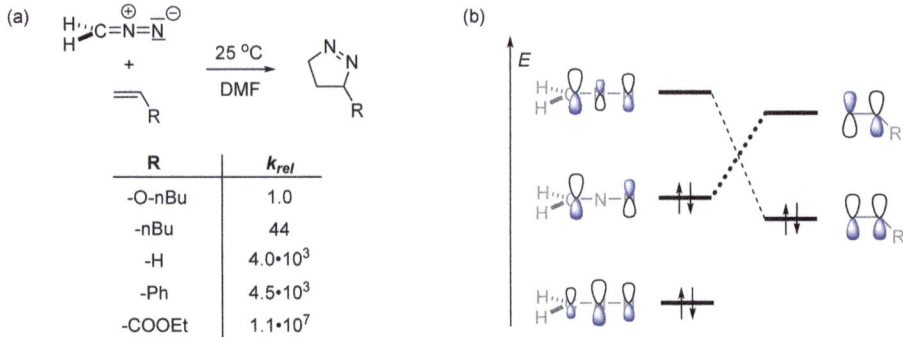

Figure 3.51 (a) Relative reaction rates for the Huisgen reaction of diazomethane with monosubstituted alkenes. (b) Dominating
HOMO(diazomethane)/LUMO(dienophile) interaction.[42]

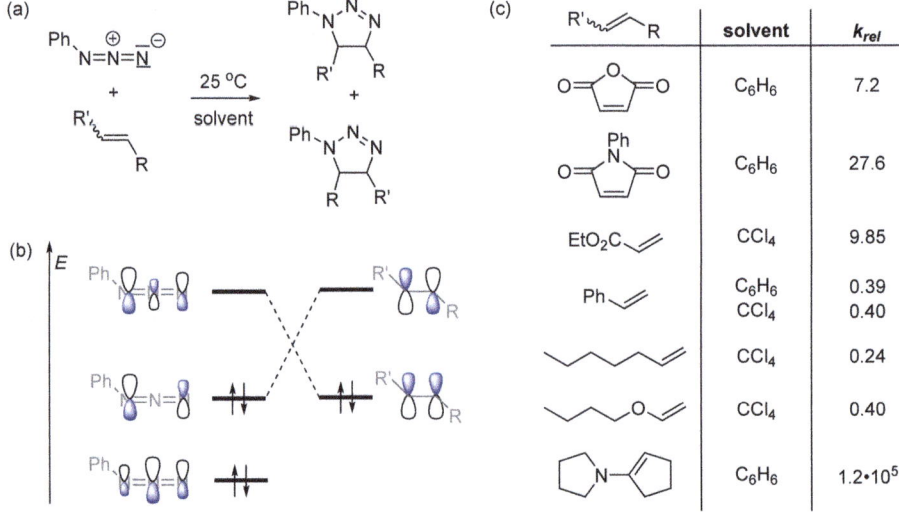

Figure 3.52 (a) Huisgen reaction of phenyl azide with donor- and acceptor-
substituted alkenes. (b) A qualitative MO diagram illustrating
the comparable importance of HOMO(phenyl azide)/LUMO(dipolarophile) and HOMO(dipolarophile)/LUMO(phenyl azide)
interactions. (c) Relative rates for the reaction shown in (a) in
benzene or tetrachloromethane.[43,44]

A larger number of Huisgen reactions belong to type II and show
rate accelerations with donor- as well as acceptor-substituted dipolarophiles. The reaction of phenyl azide (or azidobenzene) is one of
the dipoles for which substituent effects in the dipolarophiles have
been studied extensively (Figure 3.52). The fastest reactions are found
here for enamines, but not for other donor-substituted alkenes, such

as enol ethers, nor for terminal olefins, such as hept-1-ene or styrene. Reaction rates increase again on introduction of acceptor substituents, as present in acrylates, maleic anhydride, or maleimides. Although larger solvent effects are known for some Huisgen reactions, this is not so here, where reactions in benzene and tetrachloromethane appear to be practically identical.

The ozonolysis of alkenes is a typical type III Huisgen reaction (Figure 3.53). Quantitative studies are in this case complicated by the fact that the primary cycloadducts (primary ozonides) are unstable with respect to a *retro*-**Huisgen reaction** (that is, a Huisgen reaction running backwards). Taking the ozonolysis of styrene in an inert solvent such as pentane as an example, initial Huisgen reaction generates the primary ozonide as an unstable intermediate, which falls apart quickly even at −70 °C in a *retro*-Huisgen reaction to benzaldehyde oxide (sometimes also referred to as the **Criegee intermediate**) and formaldehyde. These two products then undergo a second Huisgen cycloaddition with reversed regiochemistry to yield the secondary (more stable) ozonide. Please observe that this is also an example of a **hetero-Huisgen** reaction in that the dipolarophile is not an alkene or alkyne, but an aldehyde. The fact that the reaction does indeed proceed through the carbonyl oxide (Criegee) intermediate is supported by a crossover study, where this intermediate is trapped by ^{17}O-labeled benzaldehyde. The ^{17}O label is found in the final ozonide only in the ether bridge position, in full support of the stepwise mechanism proposed by Criegee.

Most Huisgen reactions proceed in a concerted manner and are thus stereospecific with respect to alkene stereochemistry. A particularly impressive example for the quantification of stereochemical fidelity

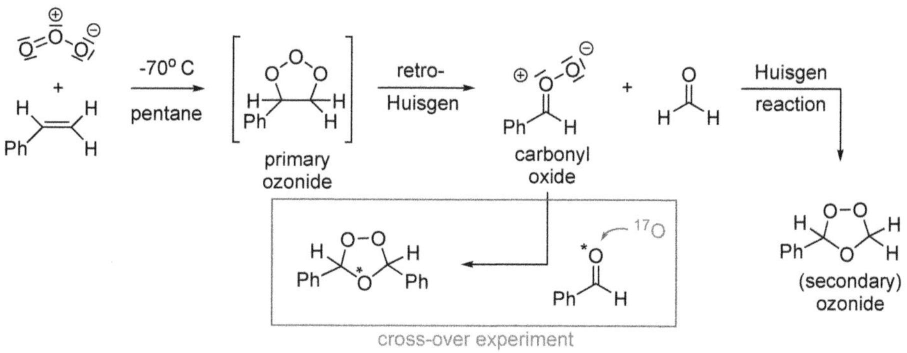

Figure 3.53 Ozonolysis of styrene in the absence and the presence of ^{17}O-labeled benzaldehyde as the trapping reagent.[45,46]

is the reaction of diazomethane with methyl (*E*)-2-methylbut-2-enoate (Figure 3.54), which has been found to provide the respective cycloadduct with more than 99.997% retention of (*E*)-stereochemistry. This, together with the finding that the dipolarophile undergoes less than 0.0006% (*E/Z*)-isomerization, puts the reaction barrier for a competing stepwise mechanism at least 25 kJ mol⁻¹ higher than the barrier for the concerted cycloaddition process.

A larger number of the dipoles shown in Figure 3.49 are unstable with respect to dimerization or unspecific decomposition, and synthetically useful Huisgen reactions of these dipoles thus combine dipole generation and subsequent cycloaddition into a one-pot procedure. Nitrile oxides, for example, can be generated under oxidative conditions from oximes, or under dehydrating conditions from nitroalkanes. The example shown in Figure 3.55 employs the latter strategy using base-catalyzed dehydration with phenylisocyanate. This is likely to involve initial attack of phenylisocyanate at the nitro group oxygen atoms, either in the neutral precursor (as shown) or in an equally possible nitroalkane anion. Subsequent intramolecular

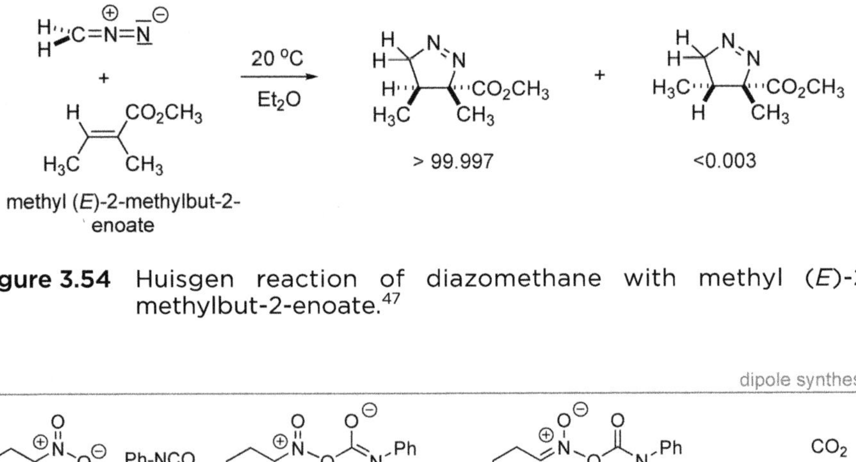

Figure 3.54 Huisgen reaction of diazomethane with methyl (*E*)-2-methylbut-2-enoate.[47]

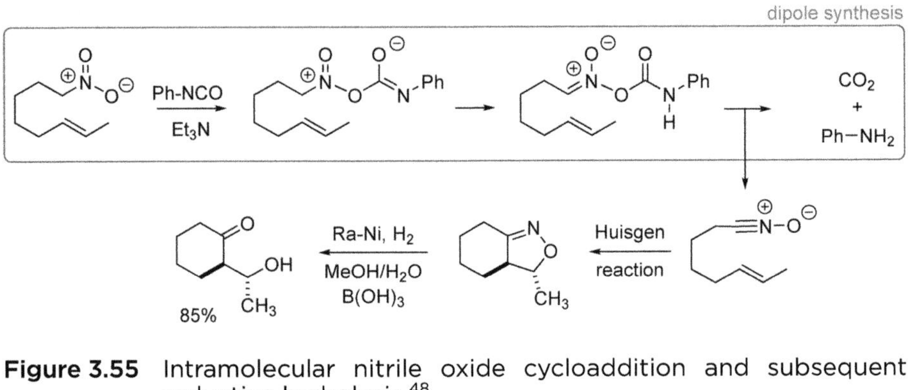

Figure 3.55 Intramolecular nitrile oxide cycloaddition and subsequent reductive hydrolysis.[48]

proton migration and elimination of CO_2 and aniline then generate the actual nitrile oxide. Intramolecular Huisgen reaction with the C–C double bond then proceeds readily and is stereospecific with respect to alkene (*E*)-stereochemistry. Subsequent reductive hydrolysis is assumed to proceed through initial (selective) reduction of the isoxazoline N–O bond, followed by hydrolysis of the C–N double bond to yield the final β-hydroxyketone in 85% yield. The perfect control of relative stereochemistry at the two chiral centers is here a consequence of the stereospecificity of the Huisgen reaction step.

Dipoles suitable for Huisgen reactions may also be formed as intermediates in a sequence of pericyclic reactions. The example shown in Figure 3.56 involves initial (inverse electron demand) hetero-Diels–Alder reaction of a 2-nitrostyrene substrate **A**, which acts as a heterodiene in its reaction with ethyl vinyl ether. Reaction with this alkene is much faster than with acrylonitrile, which is present in the reaction solution in equal amounts. The (4+2) cycloadduct **B** formed in the initial hetero-Diels–Alder reaction is capable of reacting with acrylonitrile in a follow-up Huisgen reaction. The overall sequence thus furnishes the annelated 6-5 ring system **C** built around the nitrostyrene core unit.

In recent years the copper(I)-catalyzed Huisgen reaction of azides with alkynes (which, due to its ease, is often referred to as a "click" reaction) has received considerable attention. In contrast to the reaction performed under thermal conditions, the copper(I)-catalyzed variant proceeds at much lower (often room) temperature, and with practically perfect regioselectivity. Taking the reaction of benzyl azide with phenyl propargyl ether as an example, a product ratio of 1.6 : 1.0 of the 1,4- and 1,5-disubstituted triazoles is obtained after heating the reactants for 18 h without solvent (Figure 3.57). The same compounds react at room temperature in 8 h in the presence of 1 mol% copper(II) sulfate and 5 mol% vitamin C in a mixed water/*tert*-butanol solvent system, to yield 91% of the 1,4-substituted triazole. A tentative catalytic cycle is based on monomeric or dimeric copper(I) complexes generated through initial reduction of copper(II) salts by vitamin C

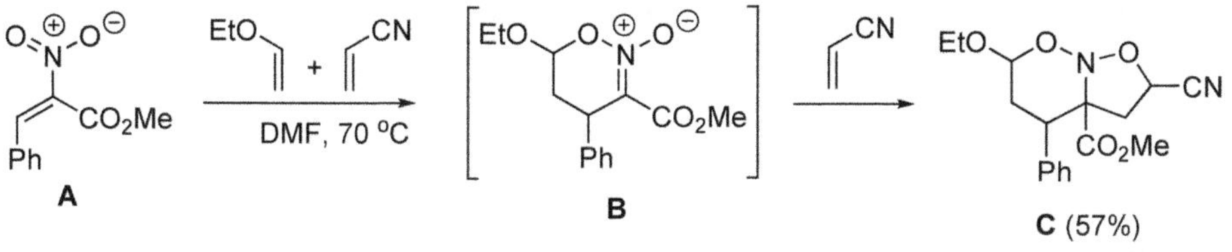

Figure 3.56 Sequential intermolecular (4+2)/(3+2) cycloaddition reactions.[49]

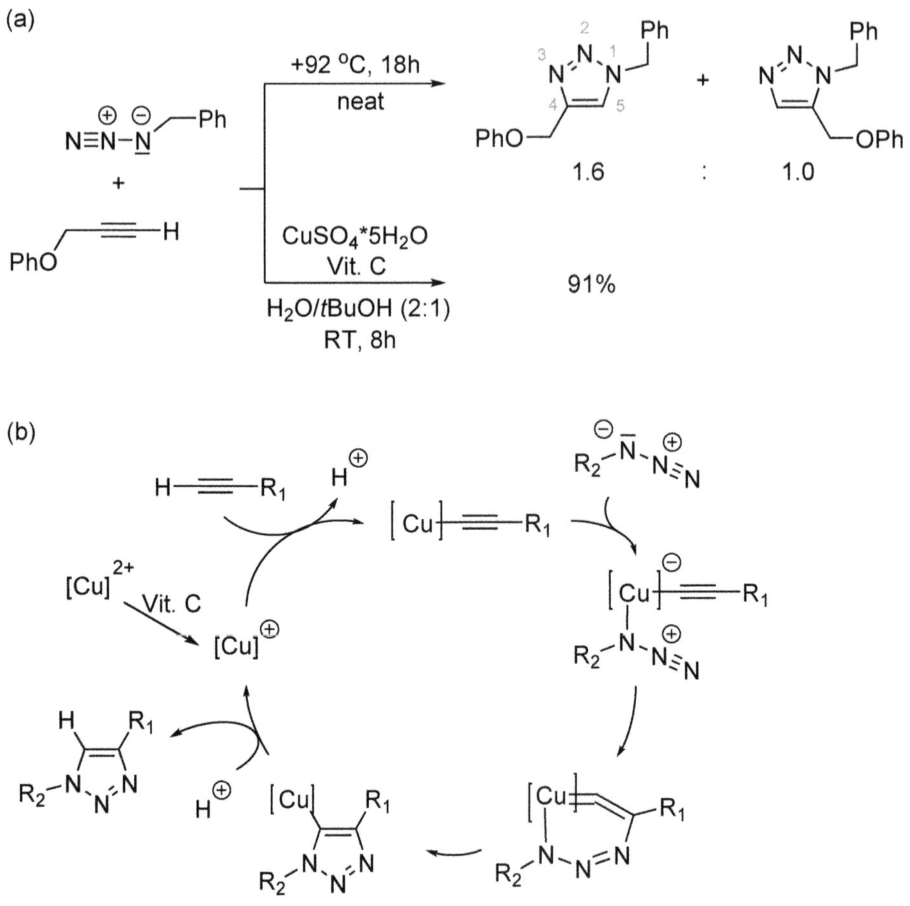

(a)

(b)

Figure 3.57 Copper(I)-catalyzed azide + alkyne "click" cycloaddition reaction.[50,51]

(Figure 3.57b). The copper(I) complexes are assumed to activate the alkyne substrate first through formation of a copper acetylide intermediate. Subsequent reaction with the azide reactant is assumed to be regiospecific with respect to copper coordination to the non-terminal nitrogen atom. This regioselectivity is retained through the follow-up cyclization/copper extrusion steps to yield the copper(I) triazolide product complex. Protonation of the copper(I) triazolide complex closes the catalytic cycle, regenerates the catalytically active copper(I) species, and generates the final triazole product. The fact that the azide "click" reaction proceeds well in aqueous solution and is compatible with molecular systems of biological relevance has led to the widespread use of this reaction for the functionalization of (oligo)nucleotides and peptides/proteins.

3.2.4 Cheletropic Reactions

The name "cheletropic reactions" refers to a group of cycloaddition reactions, where one of the reaction partners forms two new σ-bonds to the same atom. Typical examples include the reaction of carbenes with alkenes or the reaction of sulfur dioxide with 1,3-diene. Typical characteristics of this latter process can best be discussed with reference to the reaction of SO₂ with (2E,4E)-hexa-2,4-diene (Figure 3.58).

At low temperature the two reactants combine in a formal (4+1)/[4+2] cycloaddition process to the *cis*-2,5-dimethyl-3-sulfolene cycloadduct. The reaction can be driven back to the reactants at higher (>200 °C) temperatures, and the overall process can thus be employed to protect or "mask" 1,3-dienes in the context of complex synthetic sequences. The stereochemical features of the forward and the backward reactions can best be rationalized through a concerted reaction mechanism, that is suprafacial with respect to the diene reactant. The C_S-symmetric transition state for this process positions the sulfur atom at equal distances between the reacting diene carbon atoms, while the SO₂ oxygen atoms occupy *exo* and *endo* positions relative to the reacting diene. FMO analysis of this reaction is conceptually quite similar to that of other [4+2] cycloaddition processes, except that the

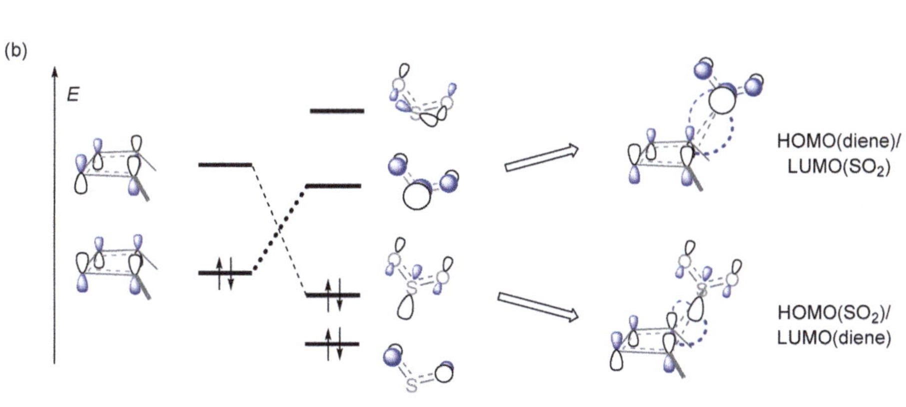

Figure 3.58 (a) Reaction of SO₂ with (2E,4E)-hexa-2,4-diene and (b) FMO analysis of this process.[52,53]

frontier orbitals of the SO_2 reactant actively involved in the process are oriented orthogonal to each other. The most relevant orbital interaction in the transition state involves the diene HOMO and the SO_2 LUMO, whose largest coefficient is located at the sulfur atom (see Figure 3.58b). Somewhat less important is the interaction of the diene LUMO with the SO_2 HOMO (mainly an sp^2-type lone-pair orbital located on sulfur). Both FMO interactions are symmetry allowed in that they allow phase-matching interactions between the carbon/sulfur centers involved in bond formation. For many substrates and reaction conditions, the reaction of SO_2 with dienes is more complex than shown in Figure 3.58. A well-known example concerns the reaction of isoprene, whose addition to SO_2 at temperatures as low as −80 °C proceeds through an acid-catalyzed (4+2)/[4+2] cycloaddition reaction to a six-membered ring sultine product. However, notably the reaction is also reversible under these conditions and already at −40 °C formation of the thermochemically more stable sulfolene product sets in (Figure 3.59).

The addition of carbenes to alkenes provides a synthetically convenient access to cyclopropanes. The actual reaction mechanism of this process depends on the electronic state of the carbene. For carbenes with a singlet ground state the addition corresponds to a concerted cycloaddition process, and the reaction of dichlorocarbene with styrene may be seen as a typical example. This carbene is conveniently generated through deprotonation of chloroform, followed by chloride elimination under phase transfer catalysis (PTC) conditions. Phase transfer catalysis refers, in this particular case, to reactions in a biphasic aqueous/organic solvent system, where PTC catalysts such as tetraalkylammonium chlorides accelerate the transport of trichloromethyl anion from the interfacial region into the organic solvent, where dichlorocarbene generation and addition to alkene substrates occur with little interference through trapping by water. Dichlorocarbene is one of the best known singlet carbenes, where the singlet ground state is characterized by a doubly occupied $C(sp^2)$-type orbital located at the central carbon atom as the carbene HOMO, and an orthogonally

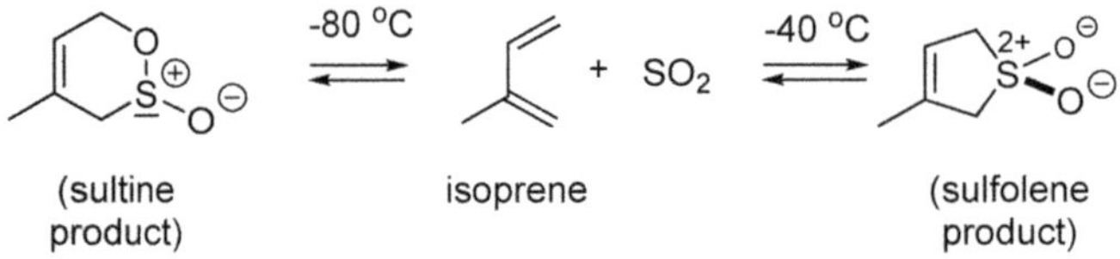

Figure 3.59   Reaction of SO_2 with isoprene.[54]

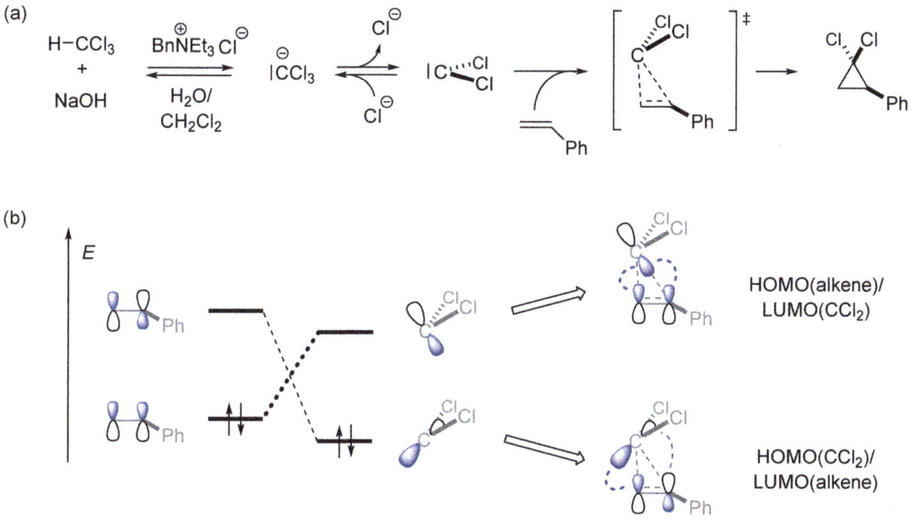

Figure 3.60 (a) Reaction of dichlorocarbene with styrene and (b) FMO analysis of this process.[55,56]

oriented p-type orbital as the overall LUMO (Figure 3.60). Addition to alkenes occurs rapidly with rather low reaction barriers through a non-symmetric transition state, where the carbene approaches the alkene double bond from one side such that the carbene Cl–C–Cl plane is tilted towards the attacked alkene face. FMO analysis of this transition-state geometry indicates dominant LUMO(CCl_2)/HOMO(alkene) interactions, while interactions between the carbene HOMO and the alkene LUMO are less relevant.

Whether carbene addition reactions to alkenes proceed in the concerted fashion shown in Figure 3.60 or follow an alternative stepwise pathway largely depends on the carbene electronic ground state. As can be seen from the singlet/triplet gap energies collected in Figure 3.61a, the singlet ground state is preferred in the presence of electron-pair donor substituents. This is easily understood as a consequence of lone-pair donation into the vacant p-type orbital located at the carbene center in singlet carbenes, and is particularly effective in diaminocarbenes, such as imidazol-2-ylidene. As shown in Figure 3.61b, this can be described alternatively through zwitterionic Lewis structures carrying a formal negative charge at the central carbene carbon atom and a formal positive charge on one of the adjacent nitrogen atoms. Carbenes with a triplet ground state are those substituted by alkyl and aryl substituents as well as methylene as the unsubstituted parent system.

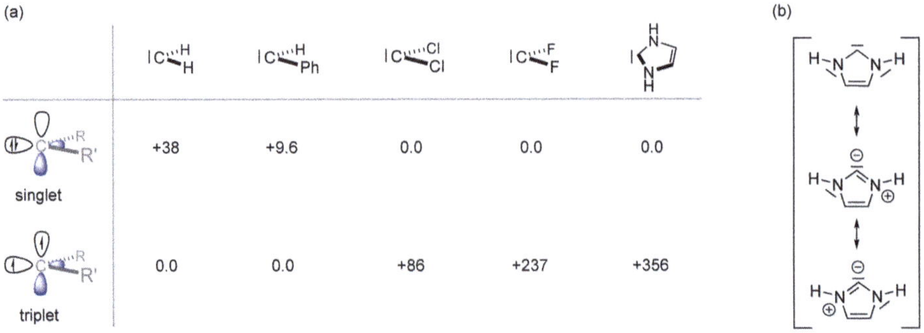

(a)   (b)

Figure 3.61 Energy differences (in kJ mol^{-1}) between singlet and triplet states for selected carbenes.[57]

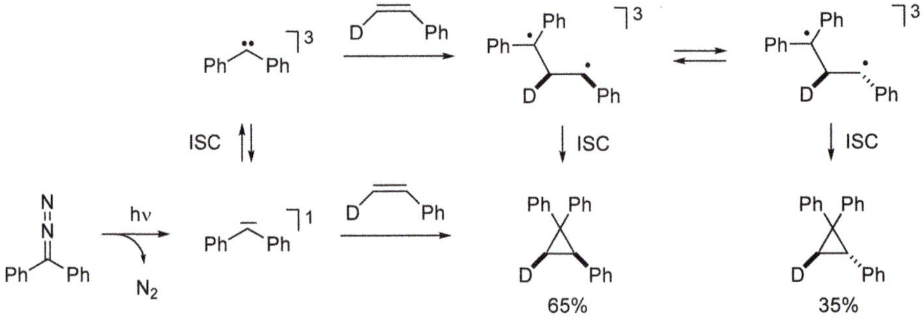

Figure 3.62 Reaction of diphenylcarbene with (Z)-β-deuteriostyrene.[57]

Reactions of triplet carbenes with alkenes occur in a stepwise manner and are therefore often accompanied by a loss of alkene stereochemistry. The reaction of diphenylcarbene with (Z)-β-deuteriostyrene is a particularly well-suited example, as the reaction proceeds with a minimum of side products (arising from C–H insertion or hydrogen abstraction reactions) to cyclopropane addition products (Figure 3.62). This experiment employs diazodiphenylmethane as a carbene precursor, whose photolysis generates singlet diphenylcarbene and N$_2$. Direct reaction of singlet diphenylcarbene with (Z)-β-deuteriostyrene leads to the cyclopropane product with retention of alkene stereochemistry. Triplet diphenylcarbene is accessible through intersystem crossing (ISC) from the singlet state, and its addition to deuteriostyrene resembles that of radical addition reactions to alkenes. The 1,3-diradical adduct formed in this addition step allows (Z)/(E)-isomerization through rotation around the former styrene double bond, as ring closure to the cyclopropane products occurs only on ISC back to the singlet state. The 65:35 mixture of *cis*/*trans*-cyclopropane products has been interpreted as the result of predominant reaction through the triplet channel. Control experiments also confirm that neither the styrene nor the products isomerize under the reaction conditions.

3.2.5 A Short Note on Carbene Chemistry Beyond Cycloaddition

The field of carbene chemistry has benefited dramatically from efforts to isolate and characterize stable singlet carbenes. A landmark achievement in this direction was reported by Arduengo and coworkers in 1991 with the X-ray crystal structure analysis of a sterically shielded diaminocarbene based on the imidazol-2-ylidene core structure (Figure 3.63a). The synthesis of this compound is based on the seemingly simple deprotonation of the respective imidazolyl chloride salt by sodium hydride in THF. In the absence of air and moisture the resulting carbene is stable enough to even allow measurement of its melting point at 240 °C. The surprising stability of this imidazol-2-ylidene carbene was thought to arise from a combination of steric shielding through its adamantyl side chains and its "aromatic" π-system containing six π-electrons. Subsequent work on diaminocarbenes lacking this characteristic have, however, also been successful, as deprotonation of 1,3-dimesitylimidazolinium chloride by potassium hydride yields the respective imidazolin-2-ylidene carbene. Again, the stability of this carbene is sufficiently high to allow X-ray crystal structure analysis, as well as determination of its melting point.

The above discoveries have sparked intense efforts to develop singlet carbenes, where the carbene center is stabilized by one or two nitrogen lone-pair donor atoms (so-called N-heterocyclic (NHC) carbenes), as Lewis base catalysts as well as ligands in transition metal catalysis. An example from the former area of application is the

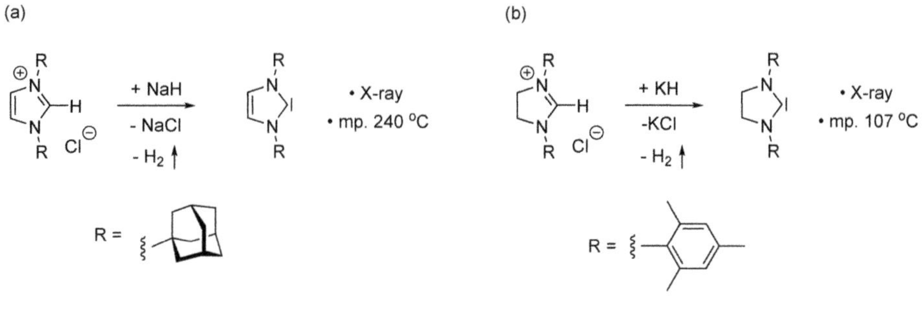

Figure 3.63 Synthesis of stable diaminocarbenes of (a) the imidazol-2-ylidene and (b) the imidazolin-2-ylidene family.[58,59]

enantioselective benzoin condensation of benzaldehyde shown in Figure 3.64. The catalytically active carbene is in this case generated through *in situ* deprotonation of a 1,2,4-triazolium salt precursor with potassium *tert*-butoxide. The resulting triazol-5-ylidene promotes the dimerization of benzaldehyde along the same mechanistic pathway as is known for cyanide. This involves initial nucleophilic attack at the carbonyl carbon atom, followed by carbon to oxygen migration of the aldehyde hydrogen atom. This generates an enol intermediate commonly referred to as a "Breslow intermediate" in honor of Ronald Breslow's efforts to unravel the mechanisms of carbene-mediated reactions. When viewed from the aldehyde reactant, the reaction corresponds to an "Umpolung" process up to this point, as the formerly electrophilic carbonyl carbon has turned into a potent nucleophile at the stage of the Breslow intermediate. Subsequent attack at a second aldehyde substrate is sufficiently selective to provide, after intramolecular proton transfer and carbene elimination, the benzoin condensation product in 83% yield and 90% enantiomeric excess (e.e.).

The mechanism shown in Figure 3.64 is biomimetic in the sense that a conceptually similar mechanism is used for substrate activation in a variety of reactions mediated by vitamin B_1 (thiamine). The pyrophosphate form of thiamine (thiamine pyrophosphate, TPP) acts as a cofactor in enzymes transforming aldehydes or 2-oxocarboxylic acid substrates, such as pyruvate. In all of these cases the catalytic cycle involves initial deprotonation of the thiazolium ring at its C2 position and formation of the corresponding thiazol-2-ylidene intermediate (Figure 3.65). This nucleophilic carbene then adds to the carbonyl

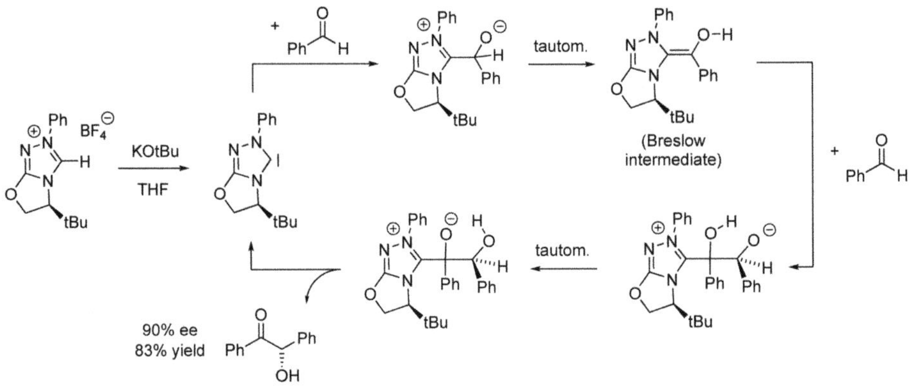

Figure 3.64 Enantioselective benzoin condensation of benzaldehyde mediated by a chiral triazol-5-ylidene catalyst.[60]

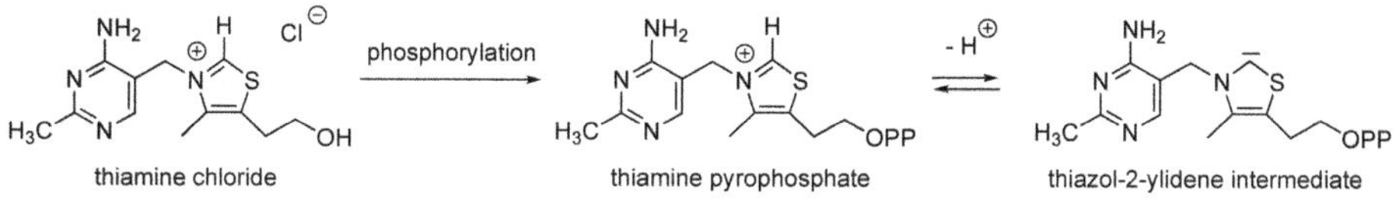

thiamine chloride   thiamine pyrophosphate   thiazol-2-ylidene intermediate

Figure 3.65 Vitamin B₁ (thiamine) and its role as a cofactor in enzymatic catalysis.

groups of the respective substrates in much the same way as shown for the triazol-5-ylidene catalyst in Figure 3.64.

3.3 Sigmatropic Rearrangements

Sigmatropic rearrangements are intramolecular pericyclic reactions and formally involve the migration of a σ-bond along at least one π-system. The particular type of sigmatropic rearrangement is defined through a pair of numbers [n,m] reflecting the final position of the σ-bond relative to its initial location. A well-known example is the [3,3]-sigmatropic rearrangement in hexa-1,5-dienes, otherwise known as the **Cope rearrangement**. As shown in Figure 3.66 the σ-bond (shown in bold) in this case formally migrates from its initial position at the left end of the two allyl fragments (labeled 1 and 1′) to the right end of the allyl fragments (labeled 3 and 3′), while the two double bonds of the allyl fragments adjust their positions accordingly. In a completely analogous way the 1,2-migration of a methyl group in carbocation intermediates (sometimes referred to as a **Wagner–Meerwein rearrangement**) can be described as a [1,2]-sigmatropic rearrangement. The migrating σ-bond is attached, in the beginning and in the end, to the methyl group carbon atom, which is reflected by the "1" in the [1,2]-designation. This is also the case in the larger number of sigmatropic rearrangements involving the migration of hydrogen atoms, as in the [1,5]-rearrangement shown in Figure 3.66. This type of nomenclature is by no means limited to pure hydrocarbon systems, as illustrated by the [2,3]-sigmatropic rearrangement of allyl sulfoxides to allyl sulfenates, which is sometimes referred to as the **Mislow rearrangement**.

From a stereochemical point of view the migration of σ-bonds in sigmatropic rearrangements may proceed with or without changing the face of the involved π-systems. For the example of the [1,5]-hydrogen migration shown in Figure 3.67a, the rearrangement is called **suprafacial** if it proceeds without a change of π-face, and **antarafacial** in the

sigmatropic rearrangement type		alternative designation

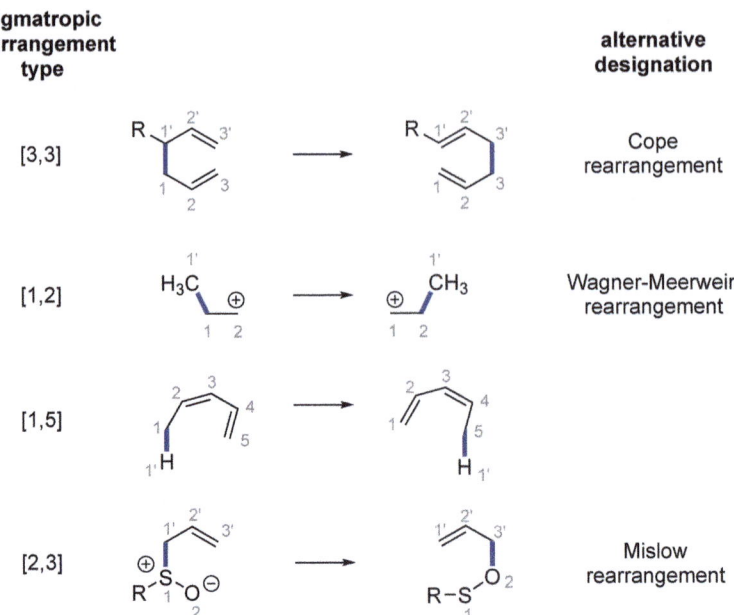

Figure 3.66   Illustrative examples for sigmatropic rearrangements.

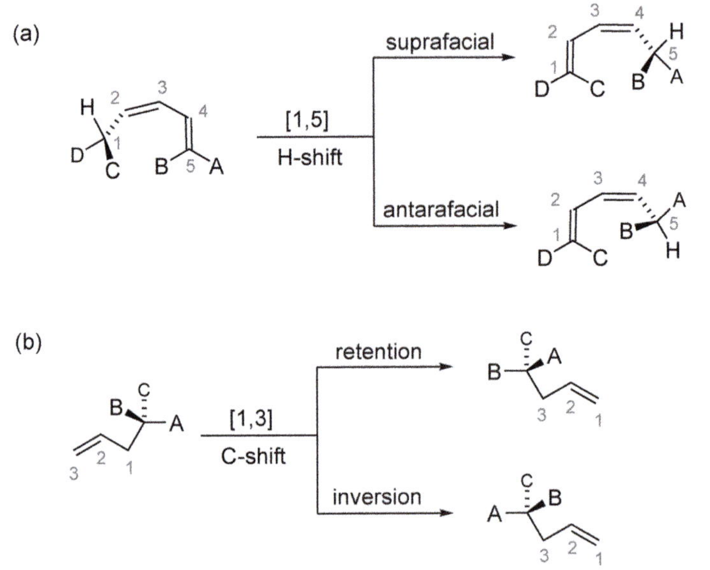

Figure 3.67   Stereochemical properties of sigmatropic rearrangements.

case of a change of face. The vast majority of experimentally known [1,5]-hydrogen shifts appear to proceed in a suprafacial manner (provided the substitution pattern allows this kind of statement to be made). In cases where the migrating σ-bond connects to a chiral center, as shown in Figure 3.67b, sigmatropic rearrangements can proceed with **inversion** or **retention** of configuration.

Following Woodward and Hoffmann the analysis of thermally allowed sigmatropic rearrangements can most easily be performed through formal homolysis of the migrating σ-bond and subsequent FMO analysis of the resulting SOMO orbitals in the perceived transition-state geometry. This is shown in Figure 3.68 for the 1,5-hydrogen migration in penta-1,3-diene, as an example of a thermally allowed suprafacial sigmatropic process. The suprafacial nature of this reaction is reflected in the transition-state geometry, where the breaking C^1–H and the forming C^5–H bonds both point downwards from the π-system. Formal homolysis of the C–H bond generates the penta-2,4-dien-1-yl radical and a hydrogen atom as the two relevant fragments. While the orbital holding the unpaired electron in the hydrogen atom is simply the hydrogen 1s atomic orbital, the SOMO of the penta-2,4-dienyl radical is the same as that already discussed in Chapter 1, Section 1.3.1. One important feature of this latter (SOMO) orbital is that the filled orbital phase points downwards at both ends. This is also the reason why FMO interactions with the 1s hydrogen orbital are constructive for the breaking C^1–H··· bond as well as the

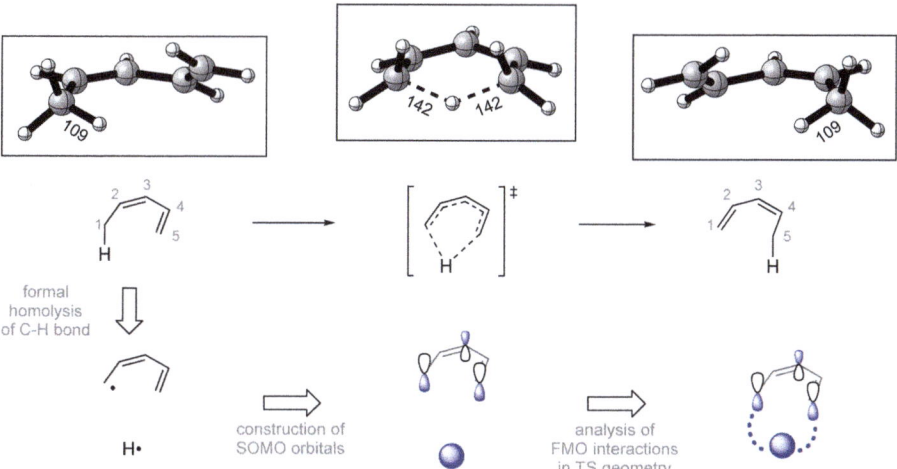

Figure 3.68 FMO analysis of the suprafacial [1,5]-hydrogen migration in penta-1,3-diene together with 3D structures of reactant, transition state, and product. Selected distances are given in pm.[61]

forming C^5–H bond. The reaction is thus allowed as a concerted pericyclic reaction under thermal conditions. Quantum mechanically calculated transition states for the 1,5-shift process show that the terminal methylene groups of the penta-2,4-dien-1-yl unit are rotated such that overlap between its π-system and the migrating hydrogen atom is maximized. At first sight this twisting is at variance with the FMO analysis, where the π-orbitals of the (planar) penta-2,4-dien-1-yl radical are employed. The impact of these skeletal deformations on the underlying MO structure are, however, quite small, and the FMO analysis as presented above is thus still valid. Variants of the above FMO analysis employ the HOMO of the migrating σ-bond and the LUMO of the diene π-system. The results obtained with this alternative choice of fragments are, however, the same as with the open-shell fragments chosen here. We will therefore continue with this latter choice for the other sigmatropic rearrangements in the following sections.

3.3.1 [3,3]-Sigmatropic Rearrangements

This class of reactions includes the **Cope** and the **Claisen** rearrangements, as two synthetically useful transformations. The stereochemical preferences of these reactions can easily be derived employing the FMO analysis described before. Assuming that the Cope rearrangement proceeds through a chair-like transition state, as shown in Figure 3.69, C1′–C1 cleavage and C3′–C3 bond formation proceeds for both allyl fragments to the same side of the π-system. Formal analysis of this transformation involves homolytic C–C bond cleavage to two allyl radicals, whose SOMO orbitals have already been discussed in

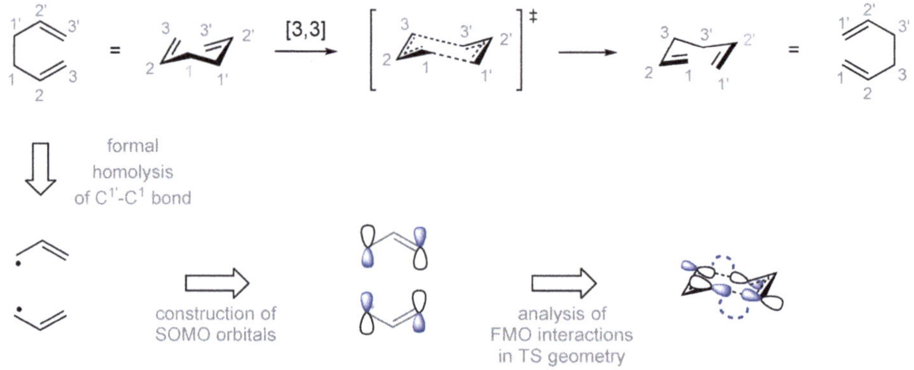

Figure 3.69 FMO analysis of the suprafacial [3,3]-sigmatropic rearrangement in hexa-1,5-diene (Cope rearrangement).

Chapter 1, Section 1.3.1. Aligning these two orbitals in the same relative orientation as in the chair transition state shows that orbital phases match up on both ends of the allyl fragments. The reaction is thus allowed as a concerted pericyclic reaction under thermal conditions.

Experimental support for a chair-like transition-state structure comes from work by Doering and Roth on the Cope rearrangement of *meso*- and *rac*-3,4-dimethylhexa-1,5-diene (Figure 3.70). The *meso*-isomer may exist in a number of different conformations, three of which are shown in Figure 3.70a as those resembling chair- and boat-like structures. The first of these conformations features one methyl substituent in a quasi-axial position and one in a quasi-equatorial position, a property that is shared with the respective chair-like transition state for the Cope rearrangement leading to (*E,Z*)-octa-2,6-diene. This product is also obtained starting from a second ground state conformation with flipped methyl group orientations (equatorial/axial instead of axial/equatorial) and represents 99.7% of the product mixture. A third pathway may be envisaged through a boat-like transition state for the Cope rearrangement with both methyl substituents in equatorial

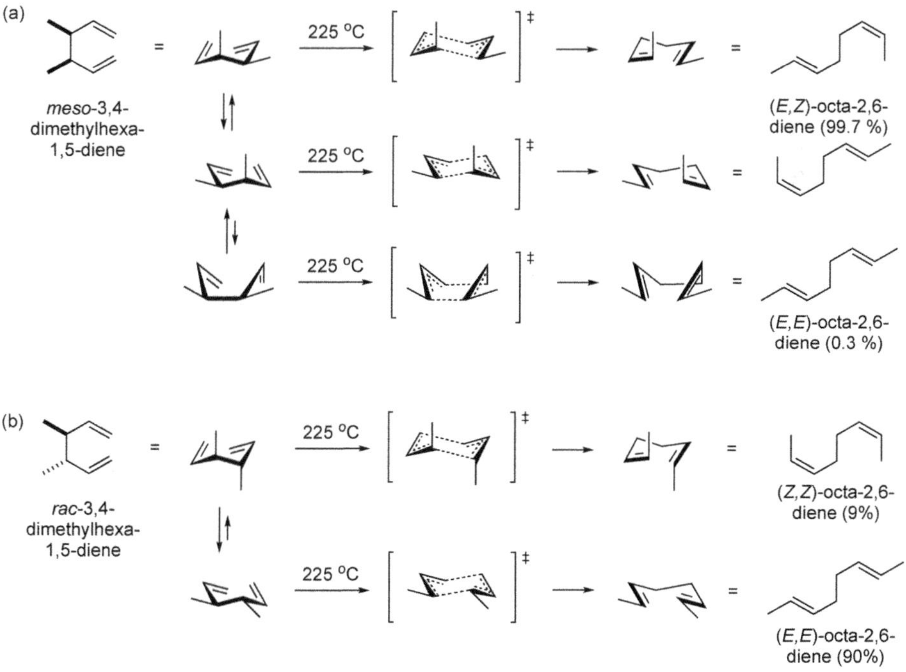

Figure 3.70 Experimental results for the Cope rearrangement of (a) *meso*- and (b) *rac*-3,4-dimethylhexa-1,5-diene.[62]

positions. The respective ground state conformation is burdened by a number of eclipsing interactions between C–H and C–C bonds, a property that is most likely also shared by the subsequent transition state. This pathway leads to (*E,E*)-octa-2,6-diene, which represents 0.3% of the product mixture. At the reaction temperature of 225 °C (or 498 K) this corresponds to a difference in reaction barriers of $\Delta G^{\ddagger}$ = $-RT\ln[k_{\text{chair}}/k_{\text{boat}}]$ = $-RT\ln[99.7/0.3]$ = 24 kJ mol^{-1} (or 5.7 kcal mol^{-1}) between the chair- and boat-like transition states. A distinctly different result is obtained for the reaction of *rac*-3,4-dimethylhexa-1,5-diene under the same reaction conditions (Figure 3.70b), whose Cope rearrangement can proceed through chair-like transition states with either bis-equatorial or bis-axial methyl group orientations. The former can safely be considered to be energetically preferred over the latter, in good agreement with the formation of (*E,E*)-octa-2,6-diene as the main reaction product (90%), while the (*Z,Z*)-isomer accounts for only 9% of the product mixture.

How substituent effects impact the mechanism of the Cope rearrangement can best be discussed with reference to the More O'Ferrall–Jencks diagram shown in Figure 3.71 for the example of

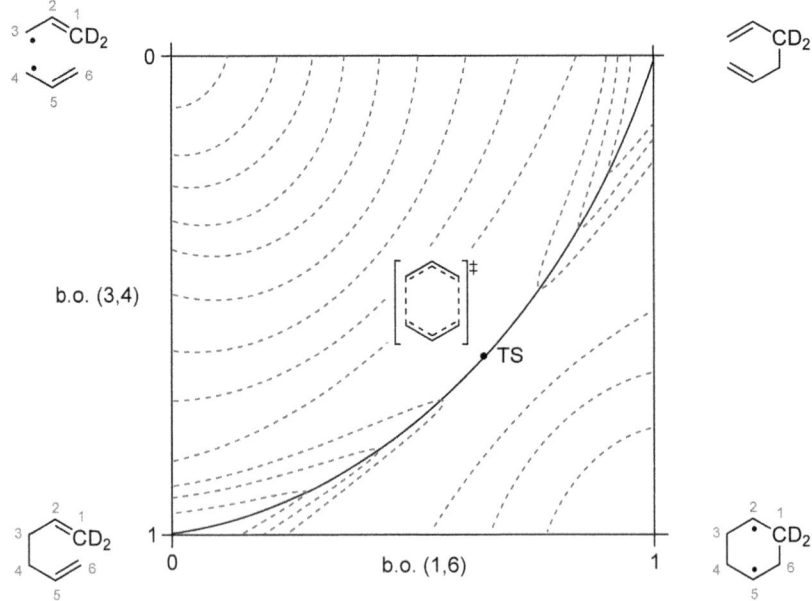

Figure 3.71 More O'Ferrall–Jencks diagram for the Cope rearrangement of 1,1-dideuteriohexa-1,5-diene. The coordinates used are the bond orders (b.o.) between carbon atoms C1 and C6 and between C3 and C4. The minimum energy reaction pathway is shown as a bold line.[63,64]

1,1-dideuteriohexa-1,5-diene. The two deuterium atoms located at the C1 position are not expected to perturb the course of the reaction notably, but serve as labels for tracking reaction progress. The lower left and upper right corners are, in this case, occupied by the reactant and product structures. In the absence of radical-stabilizing substituents the upper left structure, consisting of two allyl fragments, is energetically less favorable (at approximately +249 kJ mol^{-1} relative to the reactant) compared to the cyclohexane-1,4-diyl structure in the lower right corner. Energy estimates for this latter structure amount to 182 kJ mol^{-1}. The transition state for the concerted pathway is located 140 kJ mol^{-1} above the reactant on a shallow plateau region between the two biradical intermediates, and shares more similarity (in structural and energetic terms) with the 1,4-diyl structure. The fact that no reaction occurs through the dissociation/recombination pathway involving allyl radicals can also be gathered from the absence of products containing zero or four deuterium atoms in the temperature region used to perform the reaction. In the presence of radical-stabilizing substituents at the C3 position, the transition state moves towards the upper left corner, an effect that is easily understood considering the impact of these substituents on the allyl intermediates. A phenyl substituent at this position drops the reaction barrier from 140 to 118 kJ mol^{-1}. Following the same logic, a phenyl substituent at the C2 position moves the transition state towards the lower right corner occupied by the 1,4-diyl structure and so, in this case, the reaction barrier is reduced from 140 to 123 kJ mol^{-1}.

The anionic oxy-Cope rearrangement is one of the synthetically more useful variants of the parent Cope rearrangement (Figure 3.72). This is due to a very large acceleration of the reaction rate in the presence of negatively charged substituents at the C3 position of

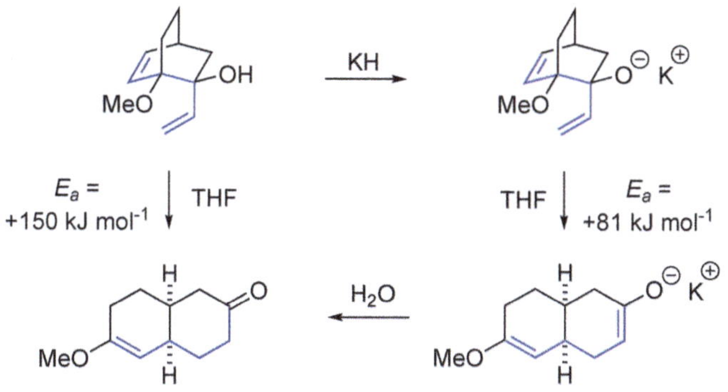

Figure 3.72 The neutral and the anionic oxy-Cope rearrangement.[65]

the parent hexa-1,5-diene framework (marked here in blue). While only moderate accelerations are observed in the presence of neutral hydroxy substituents at this position, the reaction accelerates dramatically on formation of the respective potassium alkoxide. All factors leading to essentially "free" alkoxy reactants, such as the complexation of the potassium counter-ion through crown ethers (*e.g.* 18-C-6) or through Lewis-basic solvents such as hexamethylphosphoramide (HMPA), accelerate the reaction further. Under these conditions the Cope rearrangement already proceeds efficiently at room temperature (or below). The enolate formed at the end of the reaction can be quenched by simple protonation in water or trapped with other electrophiles, such as silyl chlorides. In both cases synthetically valuable building blocks for further transformations are obtained.

The **Claisen** rearrangement is a variant of the Cope rearrangement, where one of the C3/C4 carbon atoms has been replaced by oxygen. This reaction type was first discovered by Claisen for allyl phenyl ether. Heating this substrate yields 2-allylphenol as the only product, presumably through initial [3,3]-sigmatropic rearrangement to a 6-substituted cyclohexa-2,4-diene-1-one intermediate and subsequent tautomerization to the final product (Figure 3.73a). The tautomerization step is only formally of a 1,3-sigmatropic nature, and we will see in Section 3.3.3 that the [1s,3s] variant required here is not allowed by orbital symmetry. This implies that many keto/enol tautomerization reactions are multistep reactions that often proceed under acid/base catalysis. That the initial [3,3]-sigmatropic

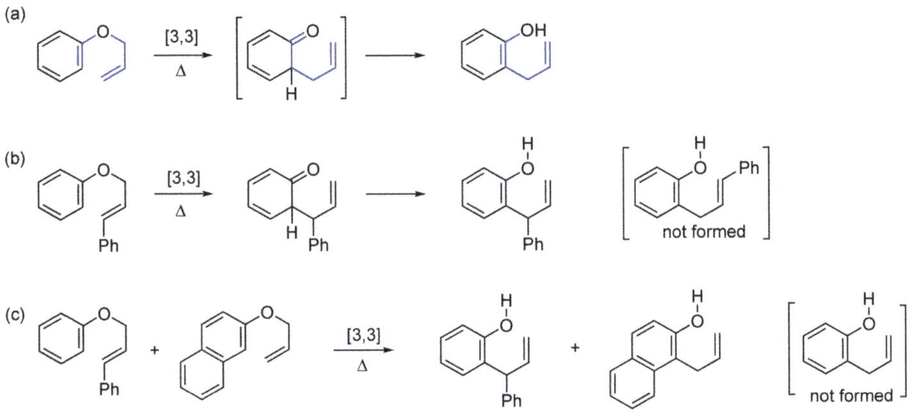

Figure 3.73   Claisen rearrangement in (a) (allyloxy)benzene; (b) (cinnamyloxy)benzene; and (c) a mixture of (cinnamyloxy)benzene and 2-(allyloxy)naphthalene.[66,67]

rearrangement shown in Figure 3.73a is best explained as a unimolecular rearrangement (and not a dissociation/recombination sequence) has been recognized early in experiments following the regiochemistry of the process (Figure 3.73b) or in crossover experiments (Figure 3.73c).

From a synthetic perspective Claisen rearrangements of aliphatic allyl vinyl ethers are far more valuable than those of aromatic substrates. Of particular importance are Claisen variants that integrate precursor synthesis and actual rearrangement into a one-pot procedure. One example for this class of reactions is the **Johnson–Claisen** rearrangement, where allylic alcohols react with orthoesters, such as ethyl orthoacetate under mildly acidic conditions (Figure 3.74). The initially formed mixed orthoester eliminates ethanol and thus generates a ketene acetal as the true substrate for the subsequent Claisen rearrangement. The stereochemistry of this latter step can be rationalized assuming a chair-like six-membered ring transition state, where substituents assume, if possible, an equatorial position. For the example shown in Figure 3.74 the reaction is highly stereoselective and yields the trisubstituted alkene product in >98% *trans* configuration.

A second synthetically valuable Claisen variant has been developed by Ireland and coworkers (often termed the **Ireland–Claisen** rearrangement) and involves silylketene acetals as substrates. These are formed in a first step from easily accessible allyl esters through deprotonation with lithium amide bases and subsequent trapping with silyl chlorides. As shown in Figure 3.75 for the example of (*E*)-crotylpropanoate, the stereoselective formation of (*E*)- or (*Z*)-enolates is possible using appropriate deprotonation conditions. Deprotonation with lithium diisopropylamide (LDA) in THF as solvent at −78 °C gives, presumably through a chair-like cyclic transition state, mainly the (*E*)-enolate, which, after trapping with *tert*-butyldimethylsilyl chloride and warming to room temperature, yields predominantly the

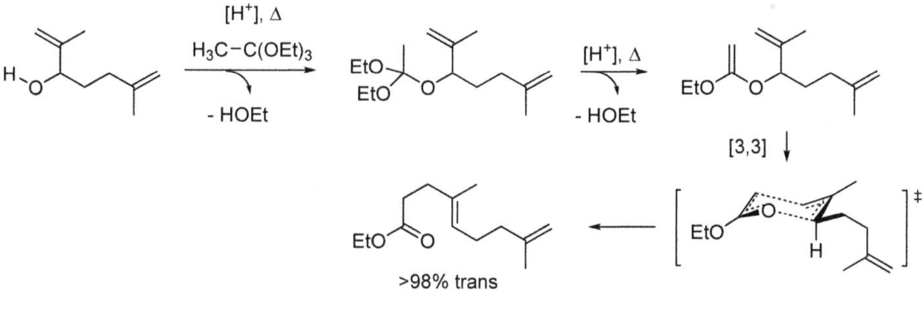

Figure 3.74 The Johnson–Claisen rearrangement.[68]

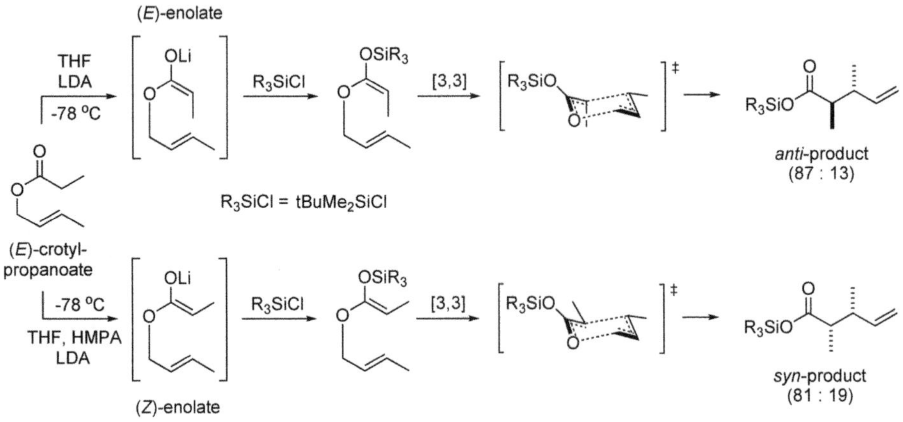

Figure 3.75 The Ireland–Claisen rearrangement of (*E*)-crotylpropanoate.[69,70]

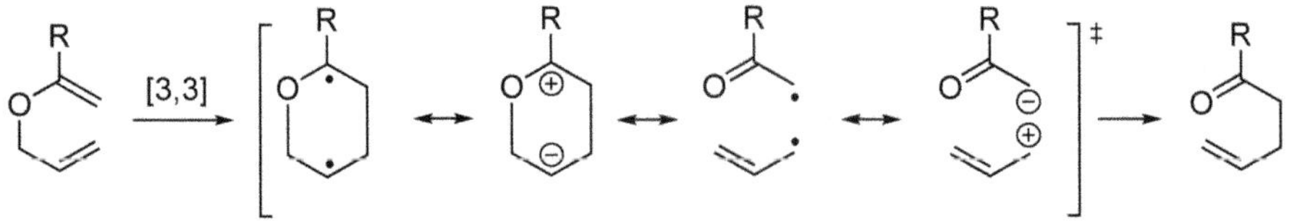

R	OCH$_3$	H	CF$_3$	CN
k_{rel}	10000	1	73	111

Figure 3.76 Substituent effects in Claisen rearrangement reactions.[71]

rearranged silylester with *anti* configuration. In contrast, addition of a Lewis-basic and thus lithium-coordinating co-solvent, such as hexamethylphosphoramide (HMPA), in the deprotonation step leads to the preferential formation of the (*Z*)-enolate and, subsequently, to the rearranged product with *syn*-stereochemistry. The actual [3,3]-sigmatropic rearrangement is assumed to proceed through a chair-like transition state in both cases.

The fact that Johnson–Claisen and Ireland–Claisen reactions work so well is also due, in part, to the accelerating effects of the donor substituents present at the C2 positions in these cases. Quantitative analysis of substituent effects at the C2 position indicates that rate acceleration is observed for both donor and acceptor substituents (Figure 3.76). Strong acceptor substituents, such as trifluoromethyl or cyano groups, accelerate the reaction by approximately two orders of magnitude, while much larger effects are observed for the methoxy group as a typical donor substituent. Efforts to rationalize these rate accelerations through stabilization of a single biradical, radical pair, or zwitterionic resonance structure at transition-state level are, however, not successful. At least for the methoxy group, rate acceleration through an increase in the reaction driving force cannot be excluded.

In contrast to the parent Cope rearrangement, solvent effects can be quite sizable for the Claisen rearrangement. This has been studied in some detail for the Claisen rearrangement of allyl vinyl ether. Reaction rates are quite similar in a number of organic solvents of variable polarity (such as pure hydrocarbons, ethers, halogenated hydrocarbons, ketones), but speed up in protic solvents, with particularly large accelerations being observed in water or water/co-solvent mixtures (Figure 3.77a). Monte Carlo solution simulations and a systematic analysis of solvent effects suggest that a combination of hydrophobic effects and hydrogen bonding to the allyl vinyl ether oxygen atom are among the controlling factors. The latter component can be visualized as shown in Figure 3.77b, where, on average, two hydrogen-bonding interactions are formed between the substrate oxygen and the protic solvent molecules at the transition-state level. The allyl vinyl ether ground state, in contrast, forms only one such interaction (of slightly reduced strength).

The Claisen rearrangement plays a role in the biosynthesis of aromatic amino acids, such as tyrosine or phenylalanine, along the shikimate pathway. One of the steps in this pathway is catalyzed by the enzyme **chorismate mutase** (CM) and involves the reaction of chorismate to prephenate (Figure 3.78a). On closer inspection we see that chorismate contains an allyl vinyl ether subunit, whose Claisen rearrangement forms the basic side chain skeleton of the final amino acid. How the enzyme accelerates this reaction by a factor $>10^6$ has been studied by co-crystallization of a fragment of the *Escherichia coli* CM with a stable molecule largely similar to the Claisen rearrangement transition state, but "stabilized" through addition of two hydrogen atoms. This compound not only acts as a competitive inhibitor of CM activity, but also serves as a "transition-state mimic"

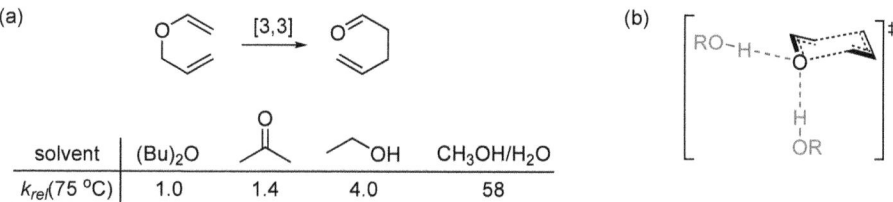

solvent	(Bu)$_2$O		^OH	CH$_3$OH/H$_2$O
k_{rel}(75 °C)	1.0	1.4	4.0	58

Figure 3.77 (a) Solvent effects in the Claisen rearrangement of allyl vinyl ether. (b) Hydrogen-bonding interactions between the corresponding chair transition state and the protic solvent.[72]

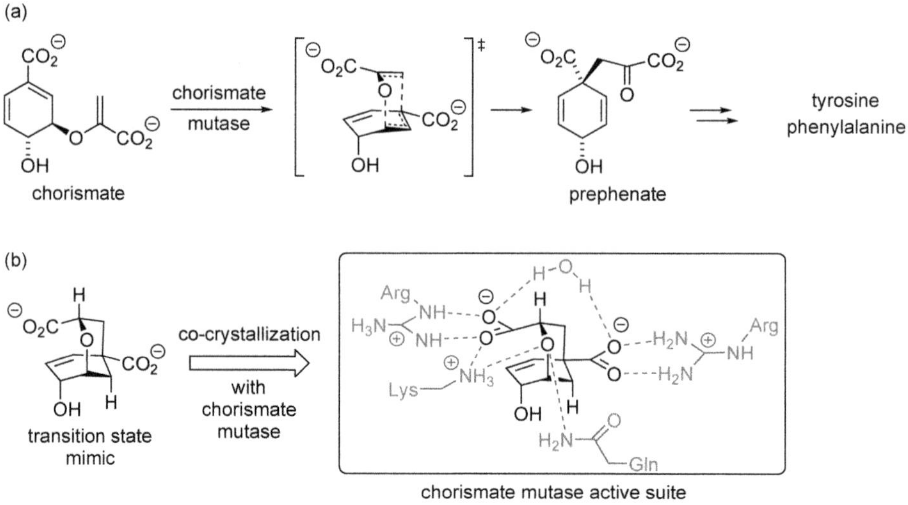

(a)

chorismate

chorismate mutase

prephenate

tyrosine
phenylalanine

(b)

transition state mimic

co-crystallization with chorismate mutase

chorismate mutase active suite

Figure 3.78 (a) Claisen rearrangement of chorismate to prephenate and (b) substrate interactions in the chorismate mutase active site.[73]

or "transition-state analog". The X-ray crystal structure of the active site region reveals binding of the dianionic mimic through two arginine salt bridges. In addition, a key lysine residue is found to make contact with one of the carboxylate groups and the oxygen atom of the substrate allyl vinyl ether subunit. A second hydrogen-bonding contact to this center is made by an adjacent glutamine side chain. In summary, these interactions suggest that enzyme activity in this case involves two major components, the hydrogen-bonding interactions to the substrate allyl vinyl ether oxygen atom and preorganization of the substrate such that reaction through a chair-like transition state proceeds efficiently. This finding has inspired synthetic efforts to design double hydrogen bond donors as small-molecule catalysts for the Claisen rearrangement, and some success in this direction has indeed been achieved with diarylureas as a compound class.

3.3.2 [1,2]-Rearrangements in Cations and Anions

[1,2]-Rearrangements count among the best and longest known sigmatropic rearrangements. This is due to the frequent occurrence of this process as a (wanted or unwanted) side reaction in carbocation chemistry. An early example for this reaction is the skeletal rearrangement that accompanies the dehydration of (+)-isoborneol to (+)-camphene, as originally discovered by Wagner and analyzed mechanistically by Meerwein (Figure 3.79). This has subsequently led to the practice of

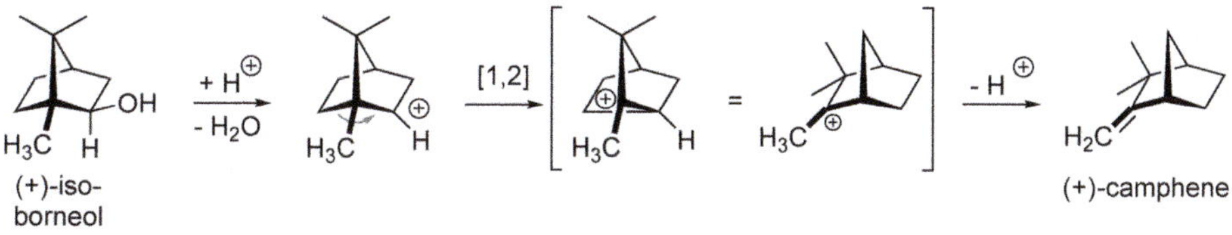

Figure 3.79 Wagner–Meerwein rearrangement in the acid-catalyzed transformation of (+)-isoborneol to (+)-camphene.[74]

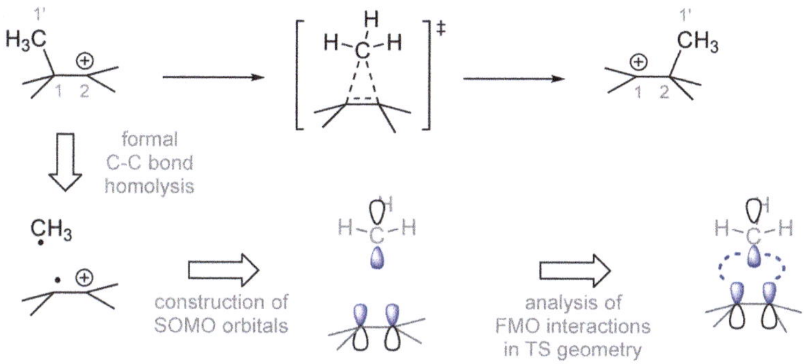

Figure 3.80 FMO analysis of the suprafacial 1,2-methyl group migration in 2,3,3-trimethylbut-2-yl cation.

naming cationic [1,2]-migrations of hydrogen atoms or alkyl groups as "**Wagner–Meerwein rearrangements**".

Reaction barriers are extremely low for many hydrogen or alkyl group [1,2]-migrations, a well-known example being the methyl group migration in 2,3,3-trimethylbut-2-yl cation shown in Figure 3.80. This process is so fast, even at −120 °C, that only a single signal is observed for all five methyl groups in the ^{1}H and ^{13}C NMR spectrum, which implies a reaction barrier below 12 kJ mol^{-1}. FMO analysis of this process following the Woodward–Hoffmann approach separates the migrating σ-bond such that a methyl radical and the radical cation of 2,3-dimethylbut-2-ene are formed. The SOMO of the former fragment is well described by a single p-type atomic orbital, while that of the latter corresponds to the HOMO of the neutral 2,3-dimethylbut-2-ene parent. Arranging these two fragments in their transition-state positions we find that orbital phases match on the bond-breaking as well as the bond-making side, and the reaction is therefore allowed with respect to the orbital symmetry rules.

Most mechanistic studies of cationic [1,2]-migration reactions are not only compatible with the one-step scenario shown in Figure 3.80, but also with a two-step mechanism where the bridged transition state turns into a true minimum. This has been studied in great detail in reactions of 2-substituted norbornanes. Taking the solvolysis of *exo*-2-norbornyl brosylate in acetic acid as an example, the exclusively formed *exo*-2-norbornyl acetate is isolated in completely racemic form (Figure 3.81). This result can be rationalized by assuming initial formation of a bridged (non-classical) cation, where the C1 and C2 positions have become fully equivalent. Backside acetate attack at either the C1 or C2 position then generates the pair of *exo*-2-norbornyl acetate enantiomers. The same final result can, of course, also be explained by assuming initial formation of a localized (classical) 2-norbornyl cation and rapid [1,2]-rearrangement to a second norbornyl cation, with the positive charge localized at the former C1 atom. Final trapping of this pair of cations with acetate then generates the racemic *exo*-2-norbornyl acetate.

Matrix isolation spectroscopic studies on structural aspects of non-classical carbocations have involved the co-deposition of *exo*-2-norbornyl chloride and SbF_5 as a strong Lewis acid at 77 K. On warming this matrix to 150 K a new infrared (IR) spectrum can be recorded that differs from that of the *exo*-2-norbornyl chloride precursor. Further warming to 200 K or higher then leads to the appearance of yet another IR spectrum that can be identified as that of 2-*exo*-norbornanol. Assignment of the IR spectrum obtained at 150 K has been aided significantly by theoretical studies of the 2-norbornyl cation with the MP2/6-31G(d) method. Geometry optimization with this latter quantum mechanical method predicts only the non-classical structure as a true minimum on the potential energy surface.

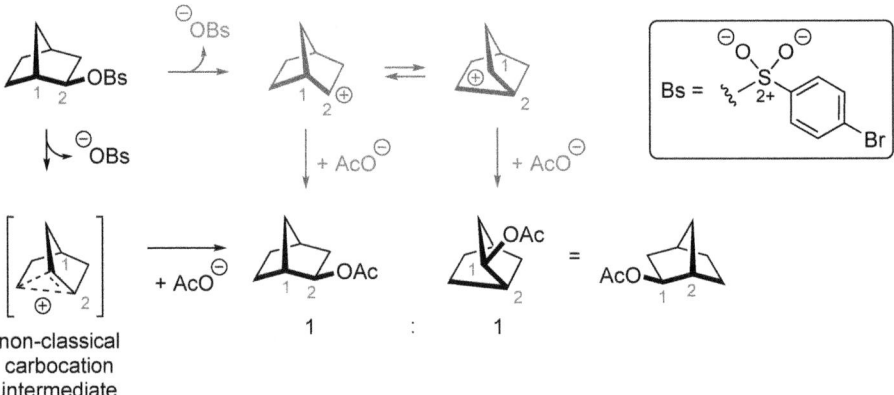

Figure 3.81   Acetolysis of *exo*-2-norbornyl brosylate.[75,76]

The theoretically calculated vibrational frequency spectrum shows, after proper scaling, diagnostic signals at 1429, 1296, and 875 cm^{-1}, which closely match the experimentally measured bands at 1435, 1305, and 878 cm^{-1}. These findings have later been confirmed in attempts to synthesize salts of 2-norbornyl cations with "inert" counter-ions and to analyze their crystal structure by X-ray diffraction. Successful efforts have involved the reaction of 2-*exo*-norbornyl bromide with AlBr$_3$ in CH$_2$Br$_2$ as the solvent at 243 K. Measurements performed at 40 K then showed three independent 2-norbornyl cations in the elementary cell, all displaying the non-classical structure with C–C distances of 180 pm on the long side and 139 pm on the short side(s) of the delocalized cation center. These structural parameters are closely matched by calculations employing the MP2/def2QZVPP method (Figure 3.82).

The **[1,2]-Wittig rearrangement** of ethers metallated in the α-position involves [1,2]-migration of an alkyl group from oxygen to the adjacent carbanionic center. Formal FMO analysis of the [1,2]-migration step again involves formal homolysis of the reacting C–O bond and construction of the SOMO orbitals of the respective fragments (Figure 3.83). The SOMO of the migrating alkyl group is identical to that involved in the cationic shift described in Figure 3.80, but the

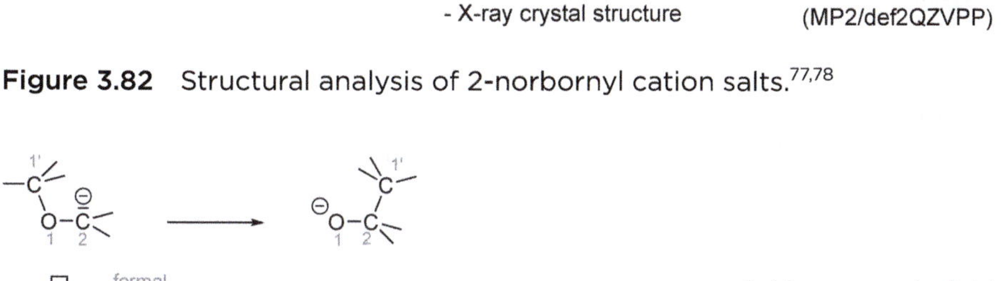

Figure 3.82 Structural analysis of 2-norbornyl cation salts.[77,78]

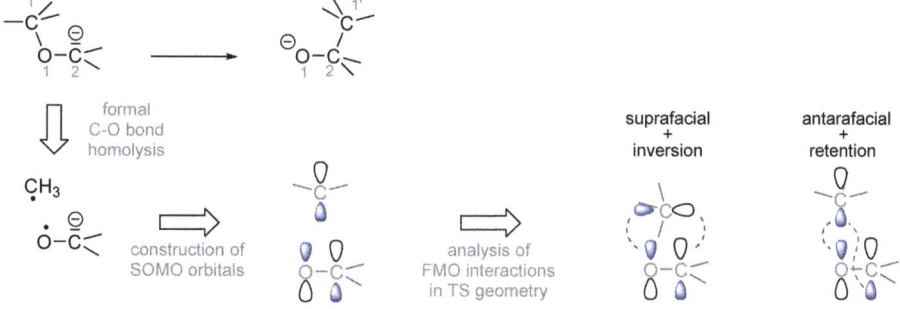

Figure 3.83 FMO analysis of the [1,2]-Wittig rearrangement.

SOMO of the ether anion fragment now equates to the formal LUMO of the respective carbonyl compound. The phase change present in this latter SOMO, on moving along the O–C bond axis, implies that the suprafacial migration of the alkyl group is no longer possible without inversion of the migrating alkyl group. This imposes severe strain onto the respective transition state due to repulsive interactions between the alkyl group substituents and the O–C bond fragment. A second alternative involves antarafacial migration of the alkyl group (without inversion) from one side of the alkoxy fragment to the other. This type of transition state is again likely to be burdened by significant strain, leaving us without a low-energy pathway for a symmetry-allowed [1,2]-migration step.

In fact, the [1,2]-Wittig rearrangement of metallated ethers proceeds quite rapidly even at low temperatures. A clear mechanistic picture has emerged from chiral substrates, such as the (*S*)-*sec*-butyl benzyl ether shown in Figure 3.84. Deprotonation (or more precisely, lithiation) of this substrate at −60 °C in THF is expected to occur at the benzylic position, as the position of highest acidity. [1,2]-Rearrangement then appears to involve formation of a radical pair intermediate, whose collapse yields the rearranged alkoxy product with (in this case) 20% retention of configuration. This implies a rather short lifetime of the radical pair intermediate, whose collapse is faster than diffusive separation of the two substrate radicals. As this type of stereochemical outcome (partial retention) is found in most experimental studies employing chiral substrates, the [1,2]-Wittig rearrangement is generally believed to follow the radical pair mechanism. This should, of course, not be confused with the formal FMO analysis depicted in Figures 3.80 and 3.83.

A larger number of reactions are known where [1,2]-rearrangement reactions occur without prior formation of a sextet intermediate. One of these reactions is the **Hock rearrangement** of benzylic hydroperoxides, which is central to the cumol process for the industrial synthesis of phenol and acetone from isopropyl benzene ("cumol"). This substrate is first oxidized to the respective hydroperoxide using air

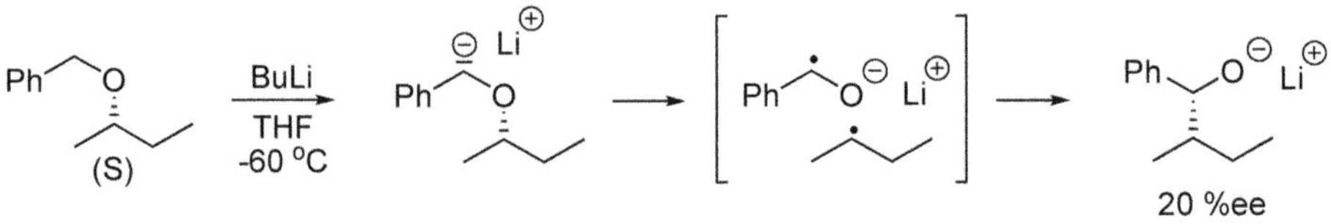

Figure 3.84 [1,2]-Wittig rearrangement in *sec*-butyl benzyl ether.[79]

as the only reagent, and subsequently rearranged to phenol and acetone under acid catalysis (Figure 3.85). This latter sequence is initiated through protonation of the terminal hydroperoxy oxygen atom.
Rather than expel water from this intermediate, form an oxenium ion
as a transient sextet species, and then migrate the phenyl substituent
to the oxygen atom, the reaction packs all of these changes into a single elementary reaction step. The structure of the transition state as
well as the migratory aptitude of the attached substituents are readily
understood on the basis of a leading FMO interaction between the
low lying σ*(O–O) orbital and the σ(C–C) orbital of the migrating C–C
bond. The strength of this donor/acceptor interaction is maximized
through *anti*-periplanar alignment of the reacting C–C and O–O
bonds (Figure 3.85).

The **Baeyer–Villiger (BV) oxidation** is a synthetically valuable reaction transforming ketones to their respective esters through reaction
with peracids. The mechanism follows a two-step sequence involving
the initial addition of the peracid to the ketone carbonyl group and
formation of the so-called "Criegee intermediate". In the second step
one of the substituents attached to the carbonyl carbon atom migrates
to the peroxy oxygen atom such that the substrate ester is formed,
together with one equivalent of acid. The intermediacy of the Criegee
intermediate is supported by labeling experiments with ^{18}O-labeled
benzophenone, where the label is found only in the benzoic acid carbonyl oxygen position (Figure 3.86). The first, and possibly also the
second, step of the BV oxidation reaction is acid catalyzed, often by
the same acid used in its peracid form as the oxidant. This implies
that the cyclic transition state shown in Figure 3.86 can be accompanied by other mechanisms involving additional molecules of acid.
Fastest reaction rates are often found for trifluoro(per)acetic acid or
other carboxylic acids carrying electron-withdrawing substituents.

The migratory aptitude of the ketone substituents is often found to
follow the order R_{tert} > cyclohexyl > R_{sek} > benzyl > Ph > R_{prim} > methyl,

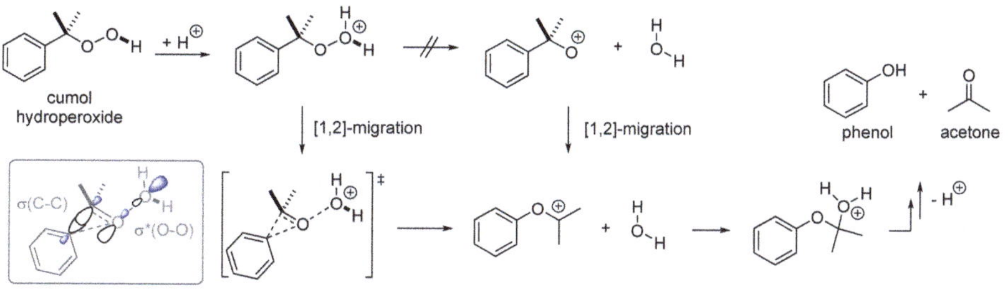

Figure 3.85 Hock rearrangement of cumol hydroperoxide.

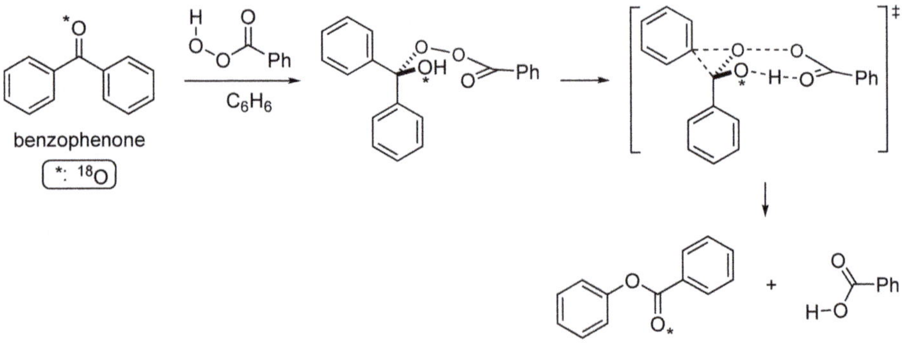

benzophenone

*: ^{18}O

Figure 3.86 Baeyer–Villiger oxidation of ^{18}O-labeled benzophenone.[80]

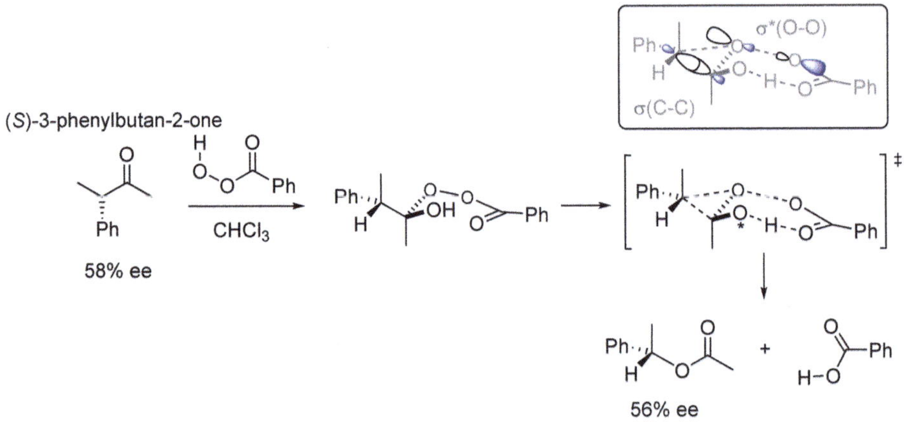

(S)-3-phenylbutan-2-one

CHCl₃

58% ee

56% ee

Figure 3.87 Baeyer–Villiger oxidation of (S)-3-phenylbutan-2-one.[81]

which implies that methyl ketones typically react to acetic acid esters. This trend is easily rationalized on the basis of the leading FMO interactions between the σ*(O–O) orbital and the σ(C–C) orbital of the migrating C–C bond (the same as in the Hock rearrangement) shown in Figure 3.87 for the Baeyer–Villiger oxidation of (S)-3-phenylbutan-2-one. In this reaction the chiral benzyl side chain migrates in preference to the methyl substituents. Comparing the optical purity of the reactant ketone and the product acetate ester we can see from this example that the migration occurs with close to perfect retention of configuration. This type of stereochemical experiment is actually quite difficult to perform due to the rapid racemization of α-chiral ketones under acidic conditions.

A third reaction featuring a combination of [1,2]-migration and nucleophilic substitution is the oxidation of boranes and boronic esters by

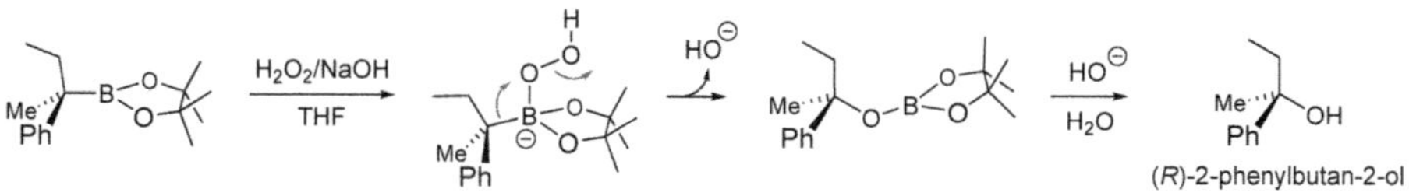

Figure 3.88 Boronic ester oxidation by peroxide under basic conditions.[82]

peroxide under basic conditions. When combined with the synthesis of boranes through hydroboration of alkenes, this represents an efficient and regioselective *anti*-Markovnikov hydration of alkenes. As shown for the chiral boronic acid pinacol ester in Figure 3.88, the reaction is initiated through nucleophilic attack of hydroperoxide anion at the boron atom and formation of a transient tetrahedral adduct. In the second step the chiral alkyl substituent migrates from boron to the adjacent oxygen atom of the peroxide unit and expels hydroxide as an anionic leaving group. The trialkylborate formed at this step is then hydrolyzed under aqueous basic conditions to yield the free alcohol, with complete retention of configuration at the migrating carbon atom. What the Hock rearrangement, the Baeyer–Villiger oxidation and the peroxide-mediated oxidation of boronates have in common is that the formation of an oxygen-centered sextet species is avoided in favor of the combination of a [1,2]-sigmatropic migration/nucleophilic substitution mechanism.

3.3.3 [1,3]-, [1,5]-, and [1,7]-Rearrangements

As already described in the introductory remarks in Section 3.3 and shown graphically in Figure 3.68, the [1,5]-sigmatropic shift of C–H bonds is symmetry allowed as a suprafacial process. As a consequence, [1,5]-hydrogen migrations occur quite readily in a number of instances and are of particular importance in five-membered ring systems. Taking the synthesis of 1-ethylcyclopentadiene as an example, the reaction of cyclopentadiene anion (prepared by deprotonation of cyclopenta-1,3-diene with an ethyl Grignard reagent) with diethyl sulfate generates 5-ethylcyclopenta-1,3-diene as the initial alkylation product (Figure 3.89). However, the [1,5]-sigmatropic hydrogen shift is rapid enough to yield 1-ethylcyclopenta-1,3-diene as the only isolated product. Due to the cyclic nature of the system one may be tempted to describe the hydrogen shift as being of a [1,2] nature. But please observe that, together with the C–H bond, the complete diene system is also involved in the rearrangement.

Hydrogen rearrangement reactions of the [1,7]-type are, in contrast, much less common, due to the more demanding stereochemical

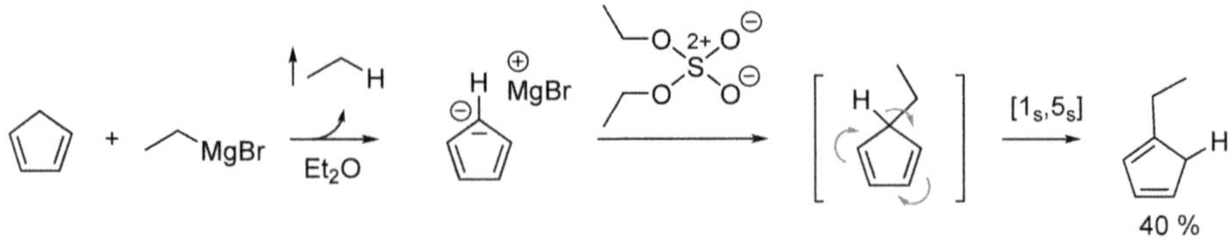

40 %

Figure 3.89 [1,5]-Sigmatropic hydrogen migration in the synthesis of 1-ethylcyclopenta-1,3-diene.[83]

requirements of this reaction type. These stereochemical requirements become readily apparent in the formal FMO analysis of this
reaction type shown in Figure 3.90a for the example of hepta-1,3,5-
triene. Formal homolytic dissection of the migrating C–H bond and
construction of the two SOMO orbitals reveals that the hepta-1,3,
5-trienyl radical SOMO structure has opposite orbital phases at the C1
and C7 termini. This implies that a concerted [1,7]-hydrogen migration is only orbital symmetry allowed as an antarafacial process.
The heptatriene system is actually large enough to allow this type
of change, from the lower to the upper face of the reacting π-system,
to occur without too much strain. In quantum mechanically calculated transition states for this reaction we can see that the hepta-1,3,
5-trienyl backbone assumes an overall helical structure in order to
maximize overlap with the migrating hydrogen atom. An example of
this type of [1,7]-hydrogen shift can be found in the biosynthesis of
vitamin D_3, where the actual triene substrate is first generated in a photochemically induced electrocyclic ring-opening reaction (see Section
3.1.2 for comparison). As shown in Figure 3.90b the [1,7]-hydrogen
shift involves an exocyclic methyl substituent as the C–H bond donor
and the bridging triene unit as the π-system relay unit, such that the
exocyclic methylene group in vitamin D_3 is generated.

The fact that the [1,7]-hydrogen migration as a concerted sigmatropic shift requires an antarafacial pathway also follows from
hydrogen migration reactions in seven-membered ring carbocycles,
such as 7-methylcyclohepta-1,3,5-triene (Figure 3.91). What might
appear to be a direct analogy to the product isomerization we have
seen in Figure 3.89 for the cyclopentadiene system, the formation
of 1-methylcyclohepta-1,3,5-triene is inhibited here due to the stereochemical demands of the antarafacial sigmatropic 1,7-hydrogen
shift. As passage of the migrating hydrogen atom through the ring
plane is not possible, the system opts for an alternative (suprafacial)
[1,5]-sigmatropic shift and rearranges to 3-methylcyclohepta-1,3,5-
triene.

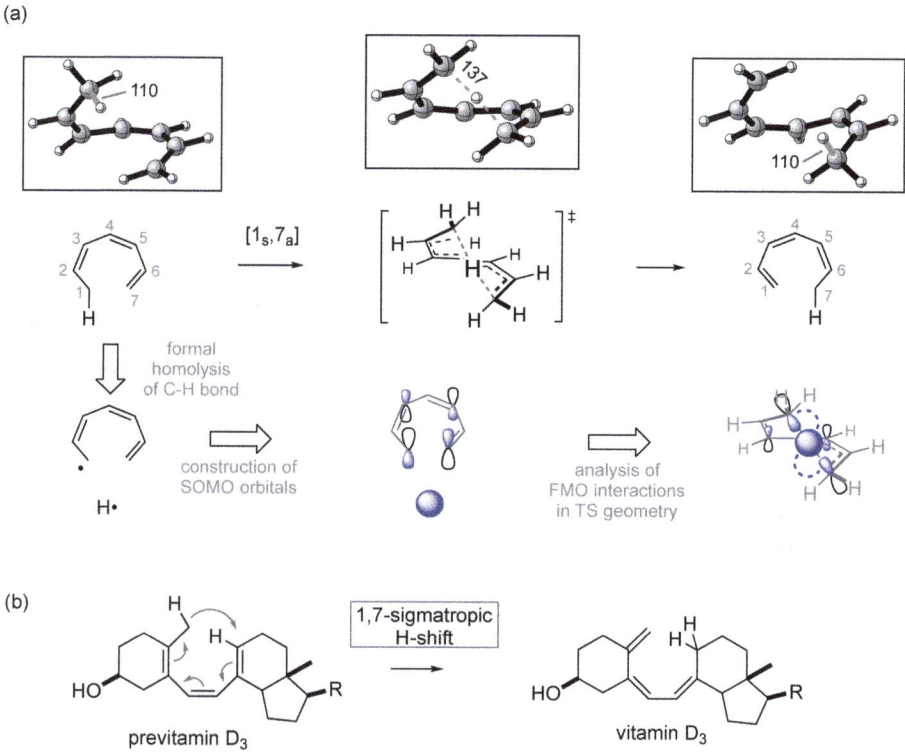

Figure 3.90 (a) FMO analysis of the antarafacial [1,7]-hydrogen migration in hepta-1,3,5-triene together with 3D structures of reactant, transition state, and product. Selected distances are given in pm. (b) [1,7]-Hydrogen migration in the synthesis of vitamin D₃.

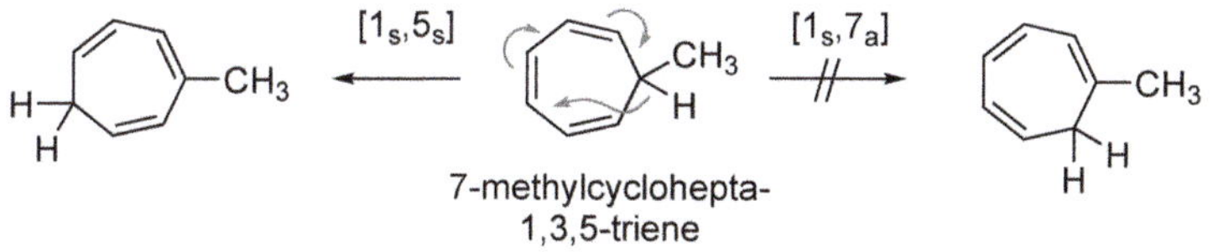

Figure 3.91 Competing [1,5]- and [1,7]-hydrogen shifts in the isomerization of 7-methylcyclohepta-1,3,5-triene.[84]

The stereochemical course of [1,3]-sigmatropic rearrangements has been studied in substituted bicyclo[3.2.0]hept-2-ene systems, such as the one shown in Figure 3.92. Thermally induced rearrangement of this compound, which carries the methyl substituent in the *exo*-orientation, occurs such that the less strained bicyclo[2.2.1]hept-2-ene (or norbornene) ring system is formed. The *exo/endo* ratio for the

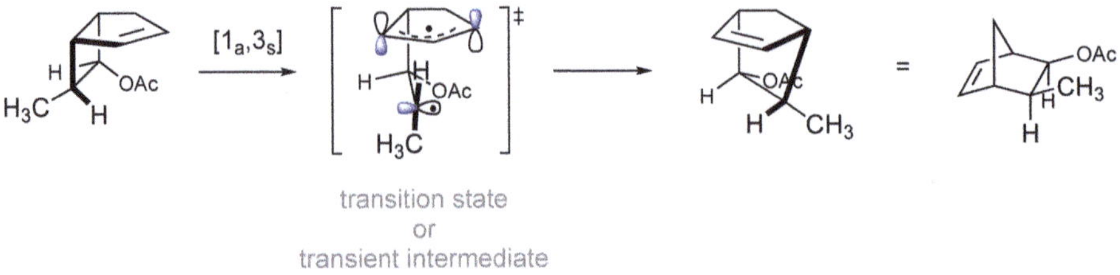

Figure 3.92 [1,3]-Sigmatropic rearrangement of a substituted bicyclo[3.2.0]hept-2-ene substrate.[85,86]

methyl group orientation in the product is 10:1, which can best be rationalized by a dominating $[1_a,3_s]$-sigmatropic process, where the carbon atom carrying the methyl substituent migrates with inversion of configuration. As indicated by the SOMO fragment structures in the proposed transition state for this transformation, this outcome is in line with the formal FMO analysis of a symmetry-allowed process. This type of conclusion has seen substantial criticism, triggered by reactive dynamics simulations by B. K. Carpenter.[86] In contrast to earlier mechanistic proposals, theoretical analysis identified a biradical intermediate instead of the biradical transition-state structure shown in Figure 3.92. More importantly, the reactive dynamics simulations show that, despite the occurrence of a true intermediate on the potential energy surface, the majority of the product is still formed with inversion of configuration at the migrating carbon atom.

Analysis of individual trajectories leading to the inverted products shows that the lifetime of the biradical intermediates is simply too short to allow for any loss of stereochemical integrity at the migrating carbon atom. This has led to the more general insight that stereochemical control of reactions is not the exclusive privilege of concerted reaction mechanisms, but may also be possible in stepwise reactions, provided that the dynamics favor one particular course of the reaction over another.

3.3.4 [2,3]-Rearrangements in Anions and Ylides

The most common [2,3]-sigmatropic rearrangement (and also the most useful from a synthetic point of view) is the [2,3]-Wittig rearrangement of metallated (or deprotonated) allyl ethers. One of the first reported examples of this reaction type is shown in Figure 3.93 and concerns the rearrangement of allyl fluorenyl ethers under basic conditions. Deprotonation of the substrate ether proceeds under comparatively mild conditions in this case due to the presence of

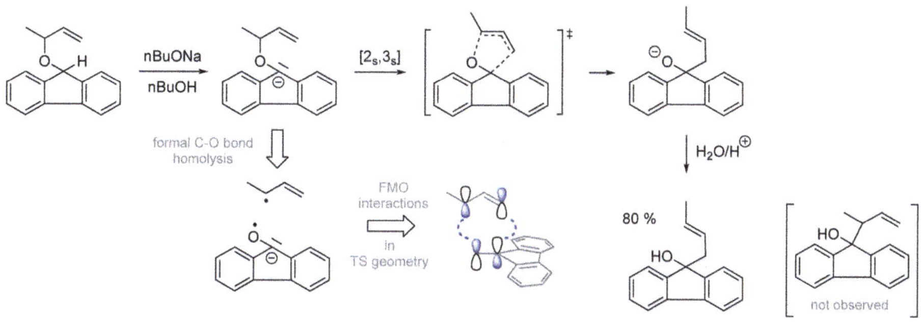

Figure 3.93 [2,3]-Wittig rearrangement of an allyl fluorenyl ether.[87]

the fluorenyl substituent. The anion formed in this case rearranges through a [2,3]-sigmatropic process such that the allyl substituent migrates to the fluorenyl C9 position. After aqueous work-up the corresponding fluorenyl alcohol is obtained in 80% yield with the newly formed allyl double bond in (*E*) orientation. According to the FMO analysis outlined in Figure 3.93, this type of rearrangement is allowed if it proceeds in a suprafacial fashion for both fragments. A doubly antarafacial variant is equally allowed in terms of orbital symmetry, but is likely to involve a very strained transition-state structure.

In many systems the [2,3]-Wittig rearrangement competes with its [1,2]-variant discussed in Section 3.3.2. While the products expected from the [1,2]-variant reaction are not observed in this particular case, small changes in the substitution pattern are known to trigger a complete change in mechanism in other systems. Most synthetic applications of the [2,3]-Wittig rearrangement involve substrates that are not quite as acidic as the fluorenyl system shown in Figure 3.93, and deprotonation of the substrate ally ethers therefore requires strong alkyl lithium bases. The α-lithioethers formed in these steps can alternatively also be generated through transmetallation from other organometallic compounds.

A particularly practical approach, sometimes referred to as the "Wittig–Still rearrangement", involves organotin ethers, as shown in the example in Figure 3.94. In this particular case the formation of the allyl ether carrying a tri-*n*-butylstannyl group next to the ether oxygen could be achieved in a stereoselective manner. Transmetallation of this type of tin compound with butyl lithium to the respective α-lithioether is known to proceed with retention of configuration and it is thus possible to follow the stereochemical course of the subsequent [2,3]-Wittig rearrangement at the lithiated carbon atom. The product stereochemistry after aqueous work-up indicates complete inversion of

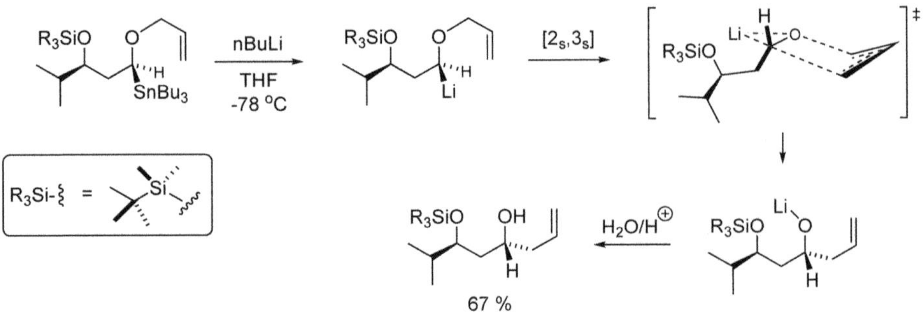

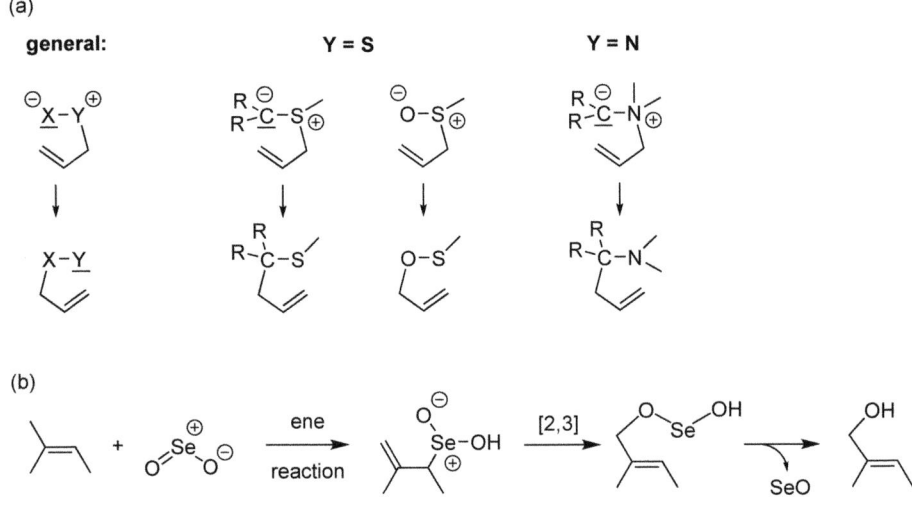

Figure 3.94 The [2,3]-Wittig rearrangement of α-stannylated allyl ethers.[88,89]

(a)

general: Y = S   Y = N

(b)

Figure 3.95 [2,3]-Sigmatropic rearrangements (a) in allyl-substituted ylides and (b) as one step in the selenium dioxide-mediated allylic oxidation of olefins.[90]

configuration in the $[2_s,3_s]$-rearrangement step. As shown in Figure 3.94 this is best rationalized with a transition state involving simultaneous suprafacial migration of the allyl fragment and C-to-O migration of the lithium cation on opposite faces of the C–O fragment.

Beyond the [2,3]-Wittig manifold of anionic (or metallated) allyl ethers, [2,3]-sigmatropic rearrangements occur in a variety of allyl-substituted ylides. As shown in the general definition in Figure 3.95a, the allyl migration occurs such that the zwitterionic ylide structure transforms into a product lacking formal charge separation. This type of reaction often integrates with the ylide-forming step into a longer reaction sequence, as exemplified by the selenium dioxide-mediated allylic oxidation of olefins in Figure 3.95b. Initial ene reaction of selenium dioxide activates one of the allylic positions of the olefin and generates an allylseleninic acid as a first transient intermediate. [2,3]-Sigmatropic rearrangement of this species then yields the selenous ester of the product (allylic) alcohol, whose ultimate formation requires elimination of selenium oxide (or possibly its solvolysis products). Although the [2,3]-sigmatropic rearrangements are shown to proceed only in the forward reaction in Figure 3.95, some of them, such as the allylsulfoxide/sulfenate rearrangement, are known to be highly reversible.

3.4 Ene Reactions

Ene reactions (sometimes also referred to as Alder-ene reactions) involve the addition of allylic substrates, commonly referred to as the "ene" component, to X–Y double and triple bonds (with X, Y = C, N, O, and S), commonly termed the "enophile". As shown in Figure 3.96 for the minimal example of propene + ethylene, this reaction derives from the related Diels–Alder reactions through formal replacement of one of the diene double bonds by a C–H bond. FMO analysis of the reaction pathway indicates that key orbital interactions in the transition state of a concerted process are those between the alkene LUMO and the propene HOMO orbitals. The latter is constructed from the π-orbital of the propene C–C double bond and the σ-orbital of the terminal propene C–H bond, such that one phase change occurs between these building blocks.

In line with the FMO interactions shown in Figure 3.96 many synthetically useful ene reactions involve enophiles having low LUMO energies, such as acceptor-substituted alkenes and alkynes. A typical

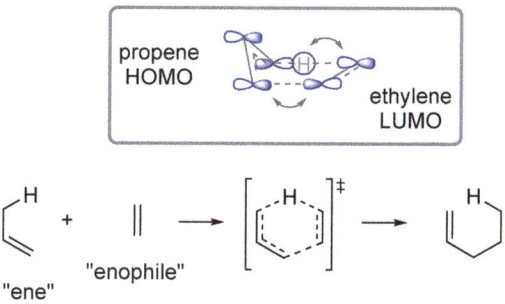

Figure 3.96 Ene reaction between propene and ethylene.

example for this case is shown in Figure 3.97 where 2-methylpropene reacts as the ene component with methyl propiolate as the enophile. On thermal activation (220 °C) this reaction produces a mixture of regioisomeric addition products. Upon complexation of the enophile with $AlCl_3$ as the Lewis acid the reaction already proceeds well at room temperature and shows improved yields and regioselectivity.

The **carbonyl ene reaction** (Figure 3.98) is a particularly useful variant where carbonyl compounds, such as aldehydes or electron-deficient ketones, act as enophiles. As already seen, the rates of reaction can also be accelerated, often dramatically, through addition of Lewis acids in this case. The carbonyl ene reaction of diethyl oxomalonate with 1-phenylcyclopentene, for example, runs to completion within 48 h at a reaction temperature of 165 °C and yields 65% of the 1:1 adduct as the only isolable (that is, low molecular weight) product. Addition of one equivalent of $SnCl_4$ as the Lewis acid gives a slightly higher yield of 71% in 12 h reaction at room temperature. Whether the catalyzed and uncatalyzed reactions actually follow the same reaction mechanism has been studied for this particular example through measurements of *para*-phenyl substituent effects on the reaction rates. Both reactions are accelerated through donor substituents in the phenyl group, but the effects are significantly larger in the Lewis acid-catalyzed reaction. Comparison of these effects to those known

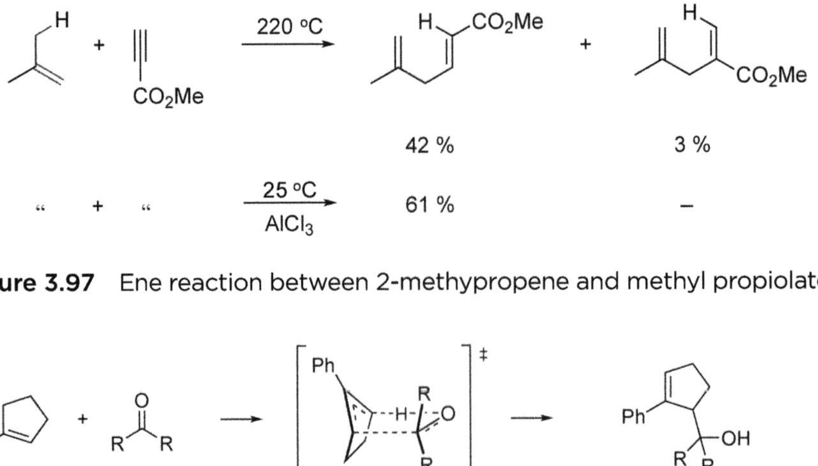

Figure 3.97   Ene reaction between 2-methypropene and methyl propiolate.[91]

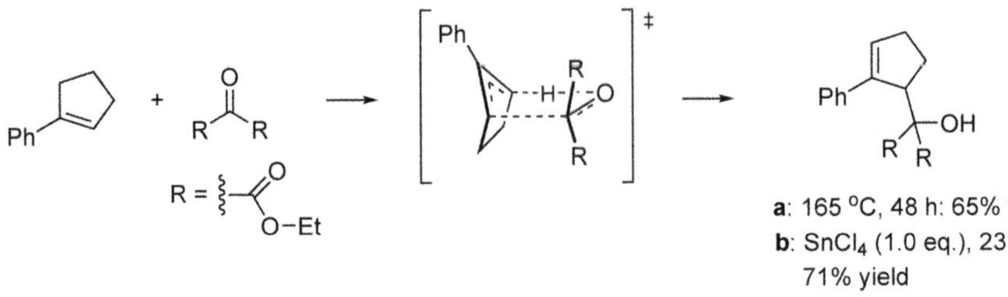

a: 165 °C, 48 h: 65% yield
b: $SnCl_4$ (1.0 eq.), 23 °C, 12 h: 71% yield

Figure 3.98   Carbonyl ene reaction between 1-phenylpentene and diethyl oxomalonate.[92]

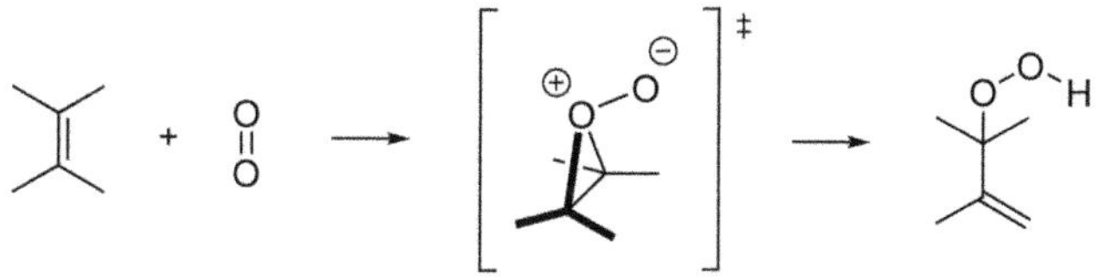

perepoxide
transition state
or
intermediate

Figure 3.99 The Schenck ene reaction of singlet oxygen ($^1O_2$) with 2, 3-dimethylbut-2-ene.[93,94]

from reactions of alkenes with carbocations suggests that the Lewis acid-catalyzed reaction may actually proceed in a stepwise manner.

A second important hetero-ene reaction involves singlet oxygen as the enophile. This reaction is sometimes also referred to as the **Schenck ene reaction** (Figure 3.99) and usually involves the photosensitized formation of singlet oxygen as the initial step. Subsequent reaction of this highly reactive species with alkyl-substituted alkenes has been proposed to involve formation of a perepoxide intermediate, whose subsequent rearrangement then yields the allyl hydroperoxide products. However, all efforts to account for the experimentally observed isotope effects in a quantitative manner are, together with results from quantum chemical studies, more in line with a "two-step no-intermediate" mechanism, where the system passes through a first and then a second transition state similar to the perepoxide structure (but with varying C–O bond distances) without encountering an intermediate minimum energy structure. This is possible on potential energy surfaces showing valley ridge inflection points between a first transition state and the final product structures (see Chapter 2, Section 2.1 for further explanation).

References

1. R. B. Woodward and R. Hoffmann, *Angew. Chem., Int. Ed. Engl.*, 1969, **8**, 781.
2. W. R. Dolbier Jr, H. Korioniak, K. N. Houk and C. Sheu, *Acc. Chem. Res.*, 1996, **29**, 471.
3. S. E. Denmark and T. K. Jones, *J. Am. Chem. Soc.*, 1982, **104**, 2642.
4. R. S. Threlkel, J. E. Bercaw, P. F. Seidler, J. M. Stryker and R. G. Bergman, *Org. Synth.*, 1987, **65**, 42.
5. E. N. Marvell, G. Caple and B. Schatz, *Tetrahedron Lett.*, 1965, 385.
6. R. B. Bates and D. A. McCombs, *Tetrahedron Lett.*, 1969, 977.
7. S. Müller and B. List, *Angew. Chem., Int. Ed.*, 2009, **48**, 9975.

8. B. Heggen, M. Patil and W. Thiel, *J. Comput. Chem.*, 2016, **37**, 280.
9. R. Huisgen, A. Dahmen and H. Huber, *J. Am. Chem. Soc.*, 1967, **89**, 7130.
10. P. v. R. Schleyer, G. W. Van Dine, U. Schöllkopf and J. Paust, *J. Am. Chem. Soc.*, 1966, **88**, 2868.
11. P. v. R. Schleyer, T. M. Su, M. Saunders and J. C. Rosenfeld, *J. Am. Chem. Soc.*, 1969, **91**, 5174.
12. R. Walsh, *Int. J. Chem. Kinet.*, 1970, **2**, 71.
13. G. Szeimies and G. Boche, *Angew. Chem., Int. Ed. Engl.*, 1971, **10**, 912.
14. P. A. Arnold and B. Carpenter, *Chem. Phys. Lett.*, 2000, **328**, 90.
15. R. R. Jones and R. G. Bergman, *J. Am. Chem. Soc.*, 1972, **94**, 660.
16. R. G. Bergman, *Acc. Chem. Res.*, 1973, **6**, 25.
17. M. Prall, A. Wittkopp and P. R. Schreiner, *J. Phys. Chem. A*, 2001, **105**, 9265.
18. C. Vavilala, N. Byrne, C. M. Kraml, D. M. Ho and R. A. Pascal Jr, *J. Am. Chem. Soc.*, 2008, **130**, 13549.
19. A. G. Myers, P. S. Dragovich and E. Y. Kuo, *J. Am. Chem. Soc.*, 1992, **114**, 9369.
20. B. Engels, C. Lennartz, M. Hanrath, M. Schmittel and M. Strittmatter, *Angew. Chem., Int. Ed. Engl.*, 1998, **37**, 1960.
21. P. Muller, Glossary of Terms used in Physical Organic Chemistry, *Pure Appl. Chem.*, 1994, **66**, 1077.
22. R. Huisgen and G. Steiner, *J. Am. Chem. Soc.*, 1973, **95**, 5054.
23. G. W. Visser, W. Verboom, D. N. Reinhoudt, S. Harkema and G. J. van Hummel, *J. Am. Chem. Soc.*, 1982, **104**, 6842.
24. H. C. Stevens, D. A. Reich, D. R. Brandt, K. R. Fountain and E. J. Gaughan, *J. Am. Chem. Soc.*, 1965, **87**, 5257.
25. E. J. Corey, Z. Arnold and J. Hutton, *Tetrahedron Lett.*, 1970, 307.
26. E. J. Corey and R. Noyori, *Tetrahedron Lett.*, 1970, 311.
27. S. Yamabe, T. Dai, T. Minato, T. Machiguchi and T. Hasegawa, *J. Am. Chem. Soc.*, 1996, **118**, 6518.
28. B. R. Ussing, C. Hang and D. A. Singleton, *J. Am. Chem. Soc.*, 2006, **128**, 7594.
29. R. Huisgen, L. A. Feiler and P. Otto, *Chem. Ber.*, 1969, **102**, 3444.
30. R. Huisgen, L. A. Feiler and G. Binsch, *Chem. Ber.*, 1969, **102**, 3460.
31. S. Yamabe, K. Kuwata and T. Minato, *Theor. Chem. Acc.*, 1999, **102**, 139.
32. K. N. Houk, Y.-T. Lin and F. K. Brown, *J. Am. Chem. Soc.*, 1986, **108**, 554.
33. H. Lischka, E. Ventura and M. Dallos, *ChemPhysChem*, 2004, **5**, 1365.
34. J. Sauer, H. Wiest and A. Mielert, *Chem. Ber.*, 1964, **97**, 3183.
35. T. Inukai and T. Kojima, *J. Org. Chem.*, 1967, **32**, 872.
36. N. J. Sisti, I. A. Motorina, M.-E. T. H. Dau, C. Riche, F. W. Fowler and D. S. Grierson, *J. Org. Chem.*, 1996, **61**, 3715.
37. J. Sauer, A. Mielert, D. Lang and D. Peter, *Chem. Ber.*, 1965, **98**, 1435.
38. J. Sauer and R. Sustmann, *Angew. Chem., Int. Ed. Engl.*, 1980, **19**, 779.
39. T. Kojima and T. Inuaki, *J. Org. Chem.*, 1970, **35**, 1342.
40. K. N. Houk, *Acc. Chem. Res.*, 1975, **8**, 361.
41. T. Inuaki and T. Kojima, *J. Org. Chem.*, 1966, **31**, 2032.
42. J. Geittner, R. Huisgen and R. Sustmann, *Tetrahedron Lett.*, 1977, **10**, 881.
43. R. Huisgen, G. Szeimies and L. Möbius, *Chem. Ber.*, 1967, **100**, 2949.
44. R. Sustmann and H. Trill, *Angew. Chem., Int. Ed. Engl.*, 1972, **11**, 838.
45. R. Criegee, *Angew. Chem., Int. Ed. Engl.*, 1975, **14**, 745.
46. C. Geletneky and S. Berger, *Eur. J. Org. Chem.*, 1998, 1625.
47. W. Bihlmaier, J. Geitner, R. Huisgen and H.-U. Reissig, *Heterocycles*, 1978, **10**, 147.
48. D. P. Curran, *J. Am. Chem. Soc.*, 1982, **104**, 4024.
49. S. E. Denmark and A. Thorarensen, *Chem. Rev.*, 1996, **96**, 137.
50. V. V. Rostovtsev, L. G. Green, V. V. Fokin and K. B. Sharpless, *Angew. Chem., Int. Ed.*, 2002, **41**, 2596.

51. C. Nolte, P. Mayer and B. F. Straub, *Angew. Chem., Int. Ed.*, 2007, **46**, 2101.
52. W. L. Mock, *J. Am. Chem. Soc.*, 1966, **88**, 2857.
53. D. Suarez, T. L. Sordo and J. A. Sordo, *J. Org. Chem.*, 1995, **60**, 2848.
54. F. Monnat, P. Vogel and J. A. Sordo, *Helv. Chim. Acta*, 2002, **85**, 712.
55. M. Fedorynski, *Chem. Rev.*, 2003, **103**, 1099.
56. A. Keating, S. M. Merrigan, D. A. Singleton and K. N. Houk, *J. Am. Chem. Soc.*, 1999, **121**, 3933.
57. S. Gronert, J. R. Keeffe and R. A. More O'Ferrall, *J. Am. Chem. Soc.*, 2011, **133**, 3381.
58. A. J. Arduengo III, R. L. Harlow and M. Kline, *J. Am. Chem. Soc.*, 1991, **113**, 361.
59. A. J. Arduengo III, J. R. Goerlich and W. J. Marshall, *J. Am. Chem. Soc.*, 1995, **117**, 11027.
60. D. Enders and T. Balensiefer, *Acc. Chem. Res.*, 2004, **37**, 534.
61. R. B. Woodward and R. Hoffmann, *J. Am. Chem. Soc.*, 1965, **87**, 2511.
62. W. von E. Doering and W. R. Roth, *Tetrahedron*, 1962, **18**, 67.
63. J. J. Gajewski, *Acc. Chem. Res.*, 1980, **13**, 142.
64. W. von E. Doering and Y. Wang, *J. Am. Chem. Soc.*, 1999, **121**, 10112.
65. D. A. Evans and A. M. Golob, *J. Am. Chem. Soc.*, 1975, **97**, 4765.
66. L. Claisen and E. Tietze, *Chem. Ber.*, 1925, 207.
67. C. D. Hurd and L. Schmerling, *J. Am. Chem. Soc.*, 1935, **59**, 107.
68. W. S. Johnson, L. Werthemann, W. R. Bartlett, T. J. Brocksom, T.-T. Li, D. J. Faulkner and M. R. Petersen, *J. Am. Chem. Soc.*, 1970, **92**, 742.
69. R. E. Ireland, R. H. Mueller and A. K. Willard, *J. Am. Chem. Soc.*, 1976, **98**, 2868.
70. X. Sun, S. L. Kenkre, J. F. Remenar, J. H. Gilchrist and D. B. Collum, *J. Am. Chem. Soc.*, 1997, **119**, 4765.
71. J. J. Gajewski, *Acc. Chem. Res.*, 1997, **30**, 219.
72. D. L. Severance and W. L. Jorgensen, *J. Am. Chem. Soc.*, 1992, **114**, 10966.
73. B. Ganem, *Angew. Chem., Int. Ed. Engl.*, 1996, **35**, 936.
74. H. Meerwein, *Justus Liebigs Ann. Chem.*, 1914, **405**, 129.
75. S. Winstein and D. S. Trifan, *J. Am. Chem. Soc.*, 1949, **71**, 2953.
76. R. A. Moss, *J. Phys. Org. Chem.*, 2014, **27**, 374.
77. F. Scholz, D. Himmel, F. W. Heinemann, P. v. R. Schleyer, K. Meyer and I. Krossing, *Science*, 2013, **341**, 62.
78. W. Koch, B. Liu, D. J. DeFrees, D. E. Sunko and H. Vancik, *Angew. Chem., Int. Ed. Engl.*, 1990, **29**, 183.
79. U. Schoellkopf, *Angew. Chem., Int. Ed. Engl.*, 1970, **9**, 763.
80. W. von E. Doering and E. Dorfman, *J. Am. Chem. Soc.*, 1953, **75**, 5595.
81. K. Mislow and J. Brenner, *J. Am. Chem. Soc.*, 1953, **75**, 2318.
82. C. Sandford and V. K. Aggarwal, *Chem. Commun.*, 2017, **53**, 5481.
83. R. Riemenschneider, *Z. Naturforsch.*, 1963, **18b**, 641.
84. K. W. Egger, *J. Am. Chem. Soc.*, 1967, **89**, 3688.
85. J. A. Berson, *Acc. Chem. Res.*, 1972, **5**, 406.
86. B. K. Carpenter, *Angew. Chem., Int. Ed. Engl.*, 1998, **110**, 3522.
87. J. Cast, T. S. Stephens and J. Holmes, *J. Chem. Soc.*, 1960, 3521.
88. W. C. Still and A. Mitra, *J. Am. Chem. Soc.*, 1978, **100**, 1927.
89. R. Hoffmann and R. Brückner, *Angew. Chem., Int. Ed. Engl.*, 1992, **31**, 647.
90. D. A. Singleton and C. Hang, *J. Org. Chem.*, 2000, **65**, 7554.
91. B. B. Snider, *Acc. Chem. Res.*, 1980, **13**, 426.
92. M. F. Salomon, S. N. Pardo and R. G. Salomon, *J. Org. Chem.*, 1984, **49**, 2446.
93. L. M. Stephenson, M. J. Grdina and M. Orfanopoulos, *Acc. Chem. Res.*, 1980, **13**, 419.
94. D. A. Singleton, C. Hang, M. J. Szymanski, M. P. Meyer, A. G. Leach, K. T. Kuwata, J. S. Chen, A. Greer, C. S. Foote and K. N. Houk, *J. Am. Chem. Soc.*, 2003, **125**, 1319.

Subject Index

Figures are in **bold**.